Hobbes's Two Sciences

*Politics, Geometry, and
the Structure of Philosophy*

MARCUS P. ADAMS

Great Clarendon Street, Oxford, OX2 6DP,
United Kingdom

Oxford University Press is a department of the University of Oxford.
It furthers the University's objective of excellence in research, scholarship,
and education by publishing worldwide. Oxford is a registered trade mark of
Oxford University Press in the UK and in certain other countries

© Marcus P. Adams 2025

The moral rights of the author have been asserted

All rights reserved. No part of this publication may be reproduced, stored in a retrieval system, transmitted, used for text and data mining, or used for training artificial intelligence, in any form or by any means, without the prior permission in writing of Oxford University Press, or as expressly permitted by law, by licence or under terms agreed with the appropriate reprographics rights organization. Enquiries concerning reproduction outside the scope of the above should be sent to the Rights Department, Oxford University Press, at the address above.

You must not circulate this work in any other form
and you must impose this same condition on any acquirer

Published in the United States of America by Oxford University Press
198 Madison Avenue, New York, NY 10016, United States of America

British Library Cataloguing in Publication Data

Data available

Library of Congress Control Number: 2024949108

ISBN 9780198924685

DOI: 10.1093/9780198924715.001.0001

Printed and bound by
CPI Group (UK) Ltd, Croydon, CR0 4YY

The manufacturer's authorised representative in the EU for product safety is
Oxford University Press España S.A. of El Parque Empresarial San Fernando de Henares, Avenida
de Castilla, 2 – 28830 Madrid (www.oup.es/en or
product.safety@oup.com). OUP España S.A. also acts as importer into Spain
of products made by the manufacturer.

To Madeleine and Hailey

Contents

Acknowledgments	vii
List of Illustrations	ix
List of Abbreviations for Citations to Thomas Hobbes's Works	xi
1. Introduction: The Unity of Hobbes's Philosophy	1
2. Prudential Knowledge from Sense and the Mechanical Mind	16
3. Scientific Knowledge from Making and the Mechanical Mind	50
4. Demonstrating Scientific Knowledge in Geometry and Civil Philosophy	95
5. Hobbesian Natural Philosophy as Mixed Mathematics	129
6. Experience in Hobbesian Civil Philosophy and Civil History	169
Works Cited	201
Index	209

Acknowledgments

I began thinking about Hobbes's philosophy in an unconventional starting place—through his work on the science of optics. My initial thoughts were shaped by trying to make sense of Hobbes's place in the history of science, which then branched out into trying to understand his vision for a systematic philosophy with parts that include geometry, natural philosophy, politics, philosophy of language, history, and more. My earliest work was guided by Peter Machamer at the University of Pittsburgh and by many fruitful interactions with Hobbes scholars around the world through the International Hobbes Association and during my time as Assistant and later Associate Editor of the journal *Hobbes Studies*. I am appreciative of Hobbes scholars in these circles for welcoming a historian and philosopher of science into their discussions.

My work on Hobbes's systematic philosophy has been influenced by many scholars. I owe great debts to Doug Jesseph, Dan Garber, and Al Martinich. My ideas have benefited greatly from their incisive criticisms and feedback over the stages of my work on Hobbes, as well as their encouragement for this project. Since the topics of this book range over many parts of Hobbes's philosophy, I would also like to acknowledge those who have influenced me through their work, through discussions at conferences, or in correspondence: Arash Abizadeh, Jeffrey Barnouw, Alex Chadwick, Jeffrey Collins, Mary Domski, Stewart Duncan, Geoff Gorham, Helen Hattab, James Lennox, Juhana Lemetti, S.A. Lloyd, José Médina, Ted McGuire, Martine Pécharman, Paolo Palmieri, Timothy Raylor, Nicholas Rescher, Patricia Springborg, Gabriella Slomp, Ed Slowik, and Kathryn Tabb. Students in graduate seminars at the University at Albany also read drafts of chapters 2–4 and offered useful feedback.

I completed a significant portion of the work for this book while holding an ACLS Fellowship (2020–2021). I am very grateful to the American Council of Learned Societies for their support of my work on Hobbes, and I am also grateful to the University at Albany Provost, Carol H. Kim, for support of funding for my Educational Leave during that academic year.

Various parts of the arguments in the present work have been presented at conferences and reading groups over the years, including the Canadian Society for History and Philosophy of Science (2017), the Eighth Margaret Dauler Wilson Philosophy Conference at SUNY Buffalo (2018), History of Science Association (2010), International Hobbes Association (2010; 2016; 2019; 2021), the International Society for the History of Philosophy of Science—HOPOS (2012; 2014), the Pittsburgh Area Philosophy Colloquium (2011), the *Scientiae*

viii ACKNOWLEDGMENTS

conference at University of Toronto (2015), the South Central Seminar in Early Modern Philosophy (2014), and the SUNY Albany/Moscow State Philosophy Conference (2015). I thank the audiences at each venue for their feedback. Some parts of chapter 4 were published in my paper "Hobbes's Laws of Nature in *Leviathan* as a Synthetic Demonstration: Thought Experiments and Knowing the Causes," *Philosophers' Imprint* 19.5 (2019): 1–23 under a Creative Commons Attribution-NonCommercial-NoDerivatives 3.0 License. Some parts of chapter 5 first appeared in my article "Hobbes on Natural Philosophy as 'True Physics' and Mixed Mathematics," *Studies in History and Philosophy of Science* 56 (2016): 43–51. For permission to use these parts, I thank Elsevier. Some other parts of chapter 5 were published in my article "Natural Philosophy, Deduction, and Geometry in the Hobbes-Boyle Debate," *Hobbes Studies* 30 (2017): 83–107. For permission to use these parts, I thank Brill.

To my colleagues in the Philosophy Department at the University at Albany, SUNY, thank you for providing encouragement to me and for creating an environment where many different approaches to philosophical method are welcomed and appreciated. I especially wish to thank Bradley Armour-Garb, who served as Chair during the entire time I worked on this book, for unfailing support of the project. At Oxford University Press, I thank Peter Momtchiloff and Tara Werger, and also two anonymous reviewers.

My family—Shannon, Madeleine, and Hailey—has done more than they could know and certainly more than can be described in words. I thank them most of all for their love and for putting up with me, gadfly that I am.

Illustrations

3.1 Representation of a Geometrical Construction in *De corpore* XIV.   83

5.1 Rotation of a Circle versus Simple Circular Motion.   146

5.2 The Axis of a Circle Moved by Circular Motion.   162

Abbreviations for Citations to Thomas Hobbes's Works

Citations to *Leviathan*

Chapters in this book cite Hobbes's *Leviathan*, published in English (1651) and in Latin (1668), with reference to pagination in the 2012 Clarendon Edition edited by Noel Malcolm and the 1651 edition.

This convention enables readers to locate citations using either the Clarendon edition or a modern edition that provides the 1651 pagination. Citations to *Leviathan* are within a set of parentheses, such as in the following example: "To know the cause of naturall Sense, is not very necessary to the business now in hand; and I have written else-where of the same at large. Nevertheless, to fill each part of my present method, I will deliver the same in this place" (Hobbes 2012, 22; 1651, 3).

Citations to Hobbes's *English Works* and *Latin Works*

Chapters in this book cite the nineteenth-century editions edited by Molesworth of Hobbes's works using the following conventions:

- Citations to Hobbes's *English Works* note the volume and page number as follows: EW I.20.
- Citations to Hobbes's *Latin Works* note the volume and page number as follows: OL I.31.

Each citation to these works occurs within a set of parentheses, such as in the following example: Later in *De corpore* I.8, Hobbes demarcates the "subject of philosophy, or the matter upon which it reflects, [as] every body of which any generation can be conceived" (EW I.10; OL I.9).

Citations to *De corpore*

Citations to Part I ("Computation or Logic") of Hobbes's work *De corpore* are cited by reference to the *English Works* and *Latin Works* editions as noted above. Where relevant differences in the text are present, they are also cited by reference to A.P. Martinich's 1981 translation of Part I.

1
Introduction: The Unity of Hobbes's Philosophy

Thomas Hobbes maintained that his philosophy constituted a unified and complete system, but in what precise sense did he think that the branches of his philosophy were unified? This question has provoked extensive scholarship over the last half-century. Not only is answering it essential to understanding Hobbes's philosophy generally, but how one answers it significantly impacts our understanding of the *Leviathan*, Hobbes's most influential work, and of the Laws of Nature, the foundation of Hobbes's civil philosophy. *Hobbes's Two Sciences: Politics, Geometry, and the Structure of Philosophy* answers this question by situating Hobbes's civil philosophy within his epistemology of scientific knowledge as constructed by humans, a form of maker's knowledge, and by demonstrating that the relationship between pure and mixed mathematics provided Hobbes with a model for thinking about relationships between geometry and natural philosophy and between civil philosophy and civil history.

1.1 Deductivism and Disunity

Scholarship on the unity of Hobbes's philosophy can be divided into two main camps, what I call the deductivist view and the disunity view. The strongest version of the deductivist view understands Hobbes's system as embracing deductive relationships *between* disciplines, such as between geometry and natural philosophy, as well as embodying Euclidean methodology *within* those disciplines by providing demonstrations from axioms. One reason for thinking Hobbes sought to emulate Euclid is John Aubrey's account in *Brief Lives*, which recounts the likely-apocryphal tale of how forty-year-old Hobbes fell "in love with geometry" when accidentally stumbling upon an open copy of Euclid's *Elements*. This love that Hobbes supposedly felt arose from a recognition that although *Elements* I.47—the proposition on the Pythagorean Theorem—did not seem intuitively certain at first glance, after tracing through its demonstration Hobbes was "demonstratively convinced of its truth" (Aubrey 1898, 332).[1]

---

[1] Martinich (1999, 85) takes a different approach and gives reasons to take Aubrey's account seriously.

Hobbes's Two Sciences. Marcus P. Adams, Oxford University Press. © Marcus P. Adams 2025.
DOI: 10.1093/9780198924715.003.0001

2 HOBBES'S TWO SCIENCES

Although Hobbes's love for geometry is clear, holding unqualifiedly that Euclid inspired either his method of demonstration within civil philosophy or other parts of philosophy, or his thinking concerning the connections between the parts of his system of philosophy overall, is difficult to maintain alongside his criticisms of Euclid. Hobbes's complaints relate to four aspects of Euclidean geometry: the nature of the objects of geometry; the nature of definitions in geometry; the nature of geometrical construction; and the rejection of axioms ("common notions"). I will briefly discuss each of these criticisms and suggest that, while Hobbes was enthralled with aspects of Euclid's geometry, he revised it significantly to align it with his materialist philosophy and with his conviction, for which he acknowledges a debt to Aristotle (see OL V.156), that knowing scientifically requires knowing through the causes, though he restricted this to knowing through the efficient causes.[2]

Regarding the nature of geometrical objects, Hobbes argues that Euclidean geometrical objects are unintelligible and thus we have no reason to think they exist in nature; for example, he states there exist no such entities as Euclidean instantaneous, indivisible points or breadthless lines in *Six Lessons* (1656a; EW VII.202) and in *De Principiis* (1666). Against thinking of points as having no parts, as Euclid's "faulty [*vitiosa*]" definition does, Hobbes defines 'point' as that which is divisible but is not "considered" to be divisible for the purposes of a geometrical demonstration (1666, 2–3). Since "conceivability"—imaginability—constrains what can count as possible objects for philosophizing, Hobbes argues that geometrical objects are properly founded in material bodies: geometrical objects are simply bodies that humans "consider" in certain ways depending upon their interests. A Hobbesian 'line' is a part of a body with breadth that humans consider as if it had no breadth. Euclid's definition for 'straight line' is "inexcusable," Hobbes taunts in *Six Lessons*, and he suggests to readers that if Euclid had "written a *Leviathan*" then they would have reviled him for this definition with "insipid jests" (EW VII.203).

Second, Hobbes asserts that Euclidean definitions are deficient because they are non-causal. Since Hobbes holds that for a discipline to count as a science (*scientia*) it must make use of causal principles in the form of definitions, in Hobbes's assessment the fundamental Euclidean definitions, such as for 'line' or 'surface', fail to be intelligible. Although Hobbes sometimes uses terms inconsistently, from this point on I use the terms 'scientia' and 'scientific knowledge' to refer to this more restrictive form of knowledge through the causes to differentiate it from knowledge (*cognitio*) that is formed from sense-based associations and fails to provide causal knowledge, though with repeated instances it can lead to prudence.[3]

---

[2] For discussion of this restriction to efficient causes, even in Hobbes's mathematics, see Jesseph (1999, 204).

[3] Hereafter I use *scientia* for this more restrictive form of knowledge, but I recognize that Hobbes is sometimes inconsistent in the terms he uses. Sometimes he clearly distinguishes *scientia* from *cognitio*, such as in *De corpore* VI.1 (OL I.58–59; EW I.66), but other times he uses *scientia* for the more general form of 'science' understood as knowledge generically. For example, he uses 'scientia' as the 1668

INTRODUCTION: THE UNITY OF HOBBES'S PHILOSOPHY   3

Broadly speaking, the subject of chapter 2 is Hobbesian *cognitio* that can lead to prudence and the subject of chapters 3 and 4 is Hobbesian *scientia*, or scientific knowledge. The Hobbesian alternative to Euclid offers generative definitions that tell the reader (or help the one who needs to remember) how to *make* the geometrical figure in question. As a result, in *De corpore* (1655) Hobbes defines 'line' as follows: "a line is made by the motion of a point" (OL I. 63; EW I.70–71). Since Euclid's definitions are not generative, Hobbes argues in *Six Lessons* (1656a) that "they ought not to be numbered among the principles of geometry" (EW VII.184). Years later in *De Principiis*, he offers similar criticisms against Euclid's definition of 'line' and cites as support Clavius's urging to think of a line as created by the motion of a point. Hobbes offers his own definition that is consistent with his earlier views: "A line is the trace that is left behind by the motion of a body, the quantity of which is not considered [*consideratur*] in a demonstration" (1666, 4–5).

Third, a criticism related to the second, geometrical demonstration according to Hobbes requires that a potential knower mechanically construct, whether in the imagination or on paper with compass and ruler, whatever the final product of the demonstration may be. To know *scientifically*, for Hobbes, is to know through the causes that one brought about oneself in the process of constructing. This requirement holds not only for geometry but also for Hobbes's second science, civil philosophy: "Geometry therefore is demonstrable for the lines and figures from which we reason are drawn and described by ourselves and civil philosophy is demonstrable because we make the commonwealth ourselves" (EW VII.184). Geometry is the primary example of knowing scientifically because of humans' unique ability to *create* the objects under consideration—imitating God, as he says in the Epistle to the Reader of *De corpore*—and in doing so possess knowledge of the causes of those objects. Possession of actual causal knowledge, not merely possible causes like Hobbes says is the case in natural philosophy (discussed in chapter 5), is the hallmark necessary condition for Hobbesian *scientia*.

Fourth, Hobbes does not ground geometry (or philosophy generally) in axioms or "common notions."[4] Although Hobbes emphasizes the need for generative

translation of 'science' from the 1651 edition of Leviathan (cf. Hobbes 2012, 72–73) and in *De corpore* II.2 to describe why language was necessary, for without it a person's "science [*scientia*] will perish with him" (OL I.12; EW I.14). When making a clear distinction among types of knowledge in *De corpore* VI.4, Hobbes also uses *cognoscere* to describe apprehension of simplest conceptions and contrasts it with *scire* to describe knowledge of the "causes of things" acquired by analysis and synthesis (Hobbes's so-called "simplest conceptions" are discussed below in chapter 4). The Molesworth translation (EW I.68) and Martinich's translation of Part I of *De corpore* (Hobbes 1981, 292–293) obscure this distinction by translating both as 'know/known'.

⁴ Hattab (2014, 477–478) agrees that Hobbes's first principles are definitions of simplest conceptions and not axioms. For discussion of Hobbes's relationship to Euclid and other figures, see chapter 3 of Jesseph (1999). Two recent examples that take Hobbes's system to be axiomatic are Deigh (1996) and Martinich (2010). Deigh (1996, 37) claims the following: "In keeping with his well-known admiration of geometry, his belief that it supplies the right model for organizing the knowledge gained in a branch of science, Hobbes represents this body of natural law as having an axiomatic structure." Martinich

4 HOBBES'S TWO SCIENCES

definitions, he does not hold that all definitions will be generative. There must be a stopping point to a definitional regress where one generative definition is defined in terms of another and so on, but Hobbes does not hold that this regress is stopped with an axiomatic foundation, arguing in *De corpore* that even Euclidean axioms require a demonstration.[5] Instead, Hobbes grounds geometry and philosophy in definitions by "explication" of what he calls our "simplest conceptions," such as BODY and MOTION. These explicatory definitions, which I argue are not merely stipulatory definitions and serve as the starting points in thought experiments used as synthetic demonstrations, form the foundation for the causal definitions that are built using them (discussed in chapter 4). The principles of construction, for Hobbes, were to be contained within the definitions of geometric objects, like the definition for 'line' mentioned already.

Nevertheless, despite these criticisms, Hobbes uses 'Elements' in the title *Elements of Law* and in his *Elementa* trilogy (Schuhmann 1998, 120 fn. 14) and encourages his readers to consult Euclid (as well as Archimedes and Apollonius) prior to starting *De corpore* Part III ("Proportions of Motions and Magnitudes"), claiming he understands his own geometry as providing only what will be useful for natural philosophy (OL I.175–176; EW I.204). Hobbes even sometimes praises Euclid. For example, when Hobbes is considering why he has included motion within his definition of 'line', he appeals to Euclid's definition of 'sphere' as an example of the same type of definition: "Euclide defines a Sphere to be a Solid Figure described by the revolution of a Semicircle, about the unmoved Diameter" (1656a, 10; EW VII.215). Although he admits that he does not seek to "defend [his own] definition by the example that of *Euclide*" (1656a, 11; EW VII.215), he clearly means to praise Euclid's incorporation of motion ("revolution") within the definition. Elsewhere, through Interlocutor A's voice, Hobbes gives similar praise in the dialogue *Examinatio et Emendatio* (hereafter *Examinatio*) to those who include "the generation of the thing" it its definition, which B concedes is something that Euclid did in the definitions of 'sphere', 'cone', and 'cylinder', though did not do in the case of 'circle' (OL IV.86–88).

Despite these praises, Hobbes's criticisms undercut and revise the Euclidean program to such an extent that it is difficult to see why some scholars have seen Hobbes as modeling his philosophy upon it, but the deductivist view holds just that. This view is well represented among Hobbes scholars and, broadly speaking, claims that Hobbes believed that his philosophy was unified because there were

---

(2010, 168–169) alludes to Hobbes's "conception of science on the geometrical model of Euclid (as he interpreted it)" and to "his Euclidean model of geometry." Boonin-Vail (1994, 22–23) sees Hobbes's praise of the use of definitions in geometry as evidence for understanding Hobbes's science more generally as following an "axiomatic method."

[5] See *De corpore* VI.13 (OL I.72; EW I.82). On this issue of the need to demonstrate even Euclidean axioms, see Leijenhorst (2002, 144ff).

deductive connections between the various parts, often starting from axioms. Representatives of this view include Martinich (2005; 2019, 39), Peters (1956), Shapin and Schaffer (1985), and Watkins (1973). Those who hold the deductivist view appeal to passages like *De corpore* VI.6 to support their claim that when Hobbes was discussing how the parts of his philosophical system overall *could* fit together he was expressing a commitment to deductivism. Hobbes does say in *De corpore* VI.6 that by beginning at first principles in first philosophy one can next move to *consideration* of geometry, and from geometry to physics. Next, he states that "after physics we come to morals," and finally he says that moral philosophy must be studied after physics because the "causes" of the passions are "in sense-experience and imagination, which are subjects of the study of physics" (Hobbes 1981, 299; OL I.64; EW I.72–73). However, given Hobbes's epistemological commitment to what I call the "perspectival principle" (discussed in chapter 3), these and other references need not be taken as indicative of a Hobbesian deductivist metaphysics. The perspectival principle permits considering parts of nature in different ways depending upon one's interests and aims, so that when one is providing a demonstration in the science of civil philosophy one can begin with consideration of human bodies and not some lower "level", as it were.

A stronger version of the deductivist view understands Hobbes as a type of reductionist. According to the reductionist version, any description of a macroscopic body ultimately reduces to a description of the microscopic bodies responsible for all phenomena. Representatives of this view include Hampton (1986) and Ryan (1970). For example, Ryan articulates the reductionist view as follows:

> Hobbes believed as firmly as one could that all behaviour, whether of animate or inanimate matter, was ultimately to be explained in terms of particulate motion: the laws governing the motions of discrete material particles were the ultimate laws of the universe, and in this sense psychology must be rooted in physiology and physiology in physics, while the social sciences, especially the technology of statecraft, must be rooted in psychology. (1970, 102–103)

Similarly, Hampton claims that according to Hobbes something is either a "fundamental physical particle" or a "compound of one of those ultimate particles" (Hampton 1986, 12). Such a claim may seem innocuous when describing a materialist like Hobbes, but Hampton's view is just as strong as Ryan's when, in outlining Hobbes's metaphysics, she claims that Hobbesian explanations will "be in terms of the operation of the fundamental physical objects in accordance with laws" (Hampton 1986, 12). Hampton links this reductionistic model that she finds in Hobbes to his "geometric" approach and believes it extends even to statements in morality.

Some deductivists hold a more attenuated view of how Euclidean geometry inspired Hobbesian method, taking it to be realized only *within* certain disciplines.

6   HOBBES'S TWO SCIENCES

For example, immediately after discussing Aubrey's tale of Hobbes's encounter with Euclid's *Elements*, Martinich asserts that the "importance of geometry on Hobbes's philosophy can hardly be exaggerated" and claims that "[a]ccording to Hobbes, natural science should take the form of geometrical demonstrations" (1999, 85). The picture of Hobbesian natural philosophy that Martinich finds is "certain, a priori, and necessary" (1999, 85). Deigh (1996; 2003) makes a similar claim regarding a different area of Hobbes's philosophy—civil philosophy—and asserts that the definition of a Law of Nature in *Leviathan* 14 plays the role of a Euclidean axiom from which Hobbes derives the Laws of Nature in *Leviathan* chapters 14 and 15. I take the criticisms of Euclid that Hobbes makes, described above, as evidence against views like these that aim to interpret Hobbes as unqualifiedly modeling his philosophy upon an axiomatic Euclidean ideal. Hobbes did seek to do philosophy *more geometrico*, but he thought that Euclidean geometry needed to be revised significantly by grounding it within his own materialist philosophy. I will not discuss the deductivist view in much more detail in the chapters that follow except to contrast it with the alternative account that I offer.[6]

The second camp—the disunity view—takes Hobbes to be enthralled by Euclid but holds that either he allowed for other methods of demonstration than demonstration from an axiom or that in practice he ultimately gave up on deductive connections between the parts of philosophy. Generally, two features of Hobbes's thought have led Hobbes scholars to this view. The first relates to something Hobbes says in *De corpore* VI.7: "Civil is connected to moral [philosophy] in such a way that it can nevertheless be detached from it" (Hobbes 1981, 301; OL I.65; EW I.73). This detachment is permissible because, in addition to learning moral philosophy from first principles, each individual could simply study the motions of their own mind and gain knowledge of the same principles. As Hobbes puts it, "[. . .] the causes of the motions of minds are [. . .] also [known] by the experience [*experientia*] of each and every person observing those motions proper to him only" (Hobbes 1981, 301; OL I.65; EW I.73). That such a detachment is permissible is difficult to explain for the deductivist view because it makes it seem that Hobbes thought that one could develop civil philosophy entirely by introspection and apart from moral philosophy, natural philosophy, and first philosophy.

A second reason scholars have been skeptical of the deductivist approach relates to how experience functions within Hobbes's philosophy. On the assumption that Hobbes understood disciplines like natural philosophy and moral philosophy to be deductively connected to one another, it is not clear how a deduction could

---

[6] Another way of attempting to account for deductive unity among the parts of Hobbes's philosophy might look to the well-known "Table" in the English edition of *Leviathan* 9. I argue against such a move elsewhere (Adams 2021, 4–5) by pointing out that if Hobbes had intended the "Table" to display the *structure* of his philosophy then he could not make sense of his use of disciplines at the terminating points of the table, like geometry, within other disciplines, like optics, that are also terminating points in the table.

work between the two because moral philosophy must add content about human passions that is not contained within—and thus not deducible from—natural philosophy. If the deductivist view is correct, what is the source of the additional information that must be added in later disciplines? For example, even though Hobbes introduces the concept 'endeavor' within natural philosophy, when it is used in moral philosophy the concepts of 'appetite' and 'aversion'—properties of *human* bodies—must be added to it so that it can be used. Malcolm raises this worry about the need to add to concepts when moving from natural philosophy to other "levels", but he ultimately takes Hobbes to have conflicting understandings of 'science', evident in what Malcolm calls Hobbes's "floundering" in his later writings on the nature of natural philosophy (2002, 147, 155; I discuss Malcolm's worries in more detail in chapter 3, section 3.4). One potential resolution that does not require us to see Hobbes as offering conflicted views of science takes sense experience as the source of this additional information, but Hobbes would not obviously be amenable to such a move because he excludes experience from philosophy in *De corpore* I.8 (OL I. 9; EW I.10–11). Another way to avoid this difficulty would be to provide evidence that Hobbes made explicit the reduction relationship between concepts like 'appetite' and 'aversion' and the concept of 'endeavor' by showing how one can reduce claims about *human* bodies to claims about microscopic bodies (as Hampton seems to think is possible). Hobbes does not appear to do this anywhere in the corpus.

The disunity view responds to worries like these by holding that regardless of what Hobbes may have claimed in some descriptions of his philosophy, in practice his philosophy was not unified. For example, Sorell (2019) sees the inclusion of experience within the science of civil philosophy as related to persuasion; even if Hobbes excludes experience from philosophy, experience makes the science of civil philosophy *persuasive* to readers. Other representatives of the disunity view include Robertson (1886), Taylor (1938), and Warrender (1957). In addition to the worries raised already for the deductivist view, some in the disunity camp have desired to free Hobbes from what appears to be a case of deriving normative claims related to the commonwealth in civil philosophy from descriptive claims related to human passions and philosophical psychology and, ultimately, claims in natural philosophy. Taylor (1938), in particular, whose debate with Warrender ignited much attention by Hobbes scholars to this issue, was motivated by such a worry. Even if Hobbes might fall prey to the so-called naturalistic fallacy, in its desire to help Hobbes escape the fallacy the disunity view ignores Hobbes's own statements that his philosophy was unified, as well as the views of many of his contemporaries that Hobbes's views in areas like metaphysics had consequences on many other aspects of philosophy such as morality and civil philosophy. In other words, the disunity view fails to take seriously Hobbes's own views as well as the assessment of Hobbes's contemporaries, like Bramhall or Mersenne, that his claims were part of an overarching system.

8 HOBBES'S TWO SCIENCES

The foregoing discussion has highlighted failures of the two dominant scholarly views of the unity of Hobbes's philosophy. Any satisfactory account of the unity of Hobbes's philosophy must meet the following desiderata:

1. Differentiate Hobbes's understanding of scientific knowledge (*scientia*) from other forms of knowledge (*cognitio*);
2. Explain how Hobbes can simultaneously exclude experience/experiments (including civil and natural history) from philosophy but also hold that they are absolutely necessary to doing philosophy (as he says, e.g., in *De corpore* I.8 and *De homine* XI.10);
3. Account *both* for Hobbes's comments about the unity of the parts of his philosophy and his apparent allowance that some parts can be done in isolation from the system; and
4. Accommodate *both* Hobbes's descriptions of the structure of his philosophy and his practice in different parts of his philosophy.

The disunity view neglects to take #3 seriously, so I will not mention it further as a contender for a satisfactory account of the unity of Hobbes's philosophy.

The deductivist view broadly looks to method in Euclidean geometry to satisfy #1; Hobbesian *scientia* on this account amounts to deductions from an axiom or axioms. I have given reasons above against thinking that Hobbes had a Euclidean axiomatic ideal in mind. Furthermore, individual views among those in the deductivist camp differ, but broadly speaking those in the deductivist camp do not have a straightforward way of answering desideratum #2. According to the deductivist view, the dialectic between, say, Hobbes and Boyle makes it seem like Boyle is defending experimental philosophy against a deductivist Hobbes. While some in the deductivist camp have addressed #4 (notably Shapin and Schaffer 1985) by looking to Hobbes's explanatory practices, Hobbes scholars have largely ignored Hobbes's explicit citational practices within his works, such as citations of causal principles from *De corpore*, alongside his invocations of experience within explanations.

My view—what I call the *maker's knowledge* view of Hobbesian philosophy—rejects the dichotomy presented by the deductivist and disunity views by charting a middle path and does not require seeing Hobbes as having conflicted views of the nature of 'science'. I outline the key components of this view in the next subsection.

1.2 The Maker's Knowledge View: Two Sciences and Mixed Mathematics

Instead of seeing Hobbes's philosophical system as embodying an axiomatic Euclidean *more geometrico* without reservation, the maker's knowledge view links

Hobbes's understanding of scientific knowledge (*scientia*) to his own view of and work in geometry. As mentioned already, Hobbes tells his readers that in his own geometry (*De corpore* Part III) he desired to do something novel; rather than repeat the work of Euclid, Archimedes, and Apollonius, he states that his aim was to offer a new geometry in the service of physics. As a result, he offers a geometry that has objects that are drawn from considerations of material bodies and incorporates locomotion within its fundamental definitions.

The maker's knowledge view articulates Hobbesian scientific knowledge as something available only to those who imitate God by making something from simple materials. Unlike the limited, suppositional knowledge available to humans concerning events in their everyday experiences, *scientia* provides knowledge of actual causes to those who are makers (more on natural philosophy below). Thus, unlike those who achieve probabilistic prudence through repeated and varied experiences (*cognitio*), such as predictions about whether the sound I hear in the night is a friend or a foe, Hobbes holds that those with *scientia* have certain knowledge because they know through the causes (this addresses desideratum #1). Such individuals need language to remember their discoveries and use language to reason syllogistically, but fundamentally they possess scientific knowledge only because they engage in a special type of internal experience—a construction of conceptions—by which they arrive at universal knowledge on the basis of a single, specially-considered conception and from which they make figures in geometry and laws and covenants in civil philosophy. Unlike prudence-based inferences about events in everyday experience, which are from *many* occurrences of some correlation to some likely future correlation, Hobbesian scientific knowledge moves from a *single* construction to universal knowledge (from one to all). This epistemic foundation lies at the root of these two Hobbesian sciences. Chapter 2 articulates how Hobbesian prudence arises from sense and how prudent humans use language in ratiocinating to form propositions and syllogisms. Chapter 3 distinguishes prudence from scientific knowledge by tracing Hobbes's account of the cognitive ability that humans have to "consider as" conceptions in different ways, depending upon their perspective and aims, and it argues that scientific knowledge is achieved when makers begin constructions from specially considered conceptions and use language to help them remember how they considered those conceptions, what they made from them, and how they made it.

Such constructions—putting together conceptions in one's mind or drawing on paper as an aid in constructing—yield objects like SQUARE in geometry or PEACE in civil philosophy. The steps along the way from start to finish will be enshrined in language in the form of definitions that supply the necessary and sufficient conditions that instruct someone how to make the thing in question. The reader of *Leviathan*, I argue, embarks on just such a synthetic demonstration starting with human bodies considered as EQUAL and progressing through various steps until making PEACE by putting all the Laws of Nature together at one

10 HOBBES'S TWO SCIENCES

time. This account of Hobbesian demonstration is the focus of chapter 4. I argue that putting the Laws of Nature together in the science of civil philosophy to make PEACE is the same constructive act as putting together the parts of a geometric figure like SQUARE.

The role of experience in Hobbes's systematic philosophy relates to desiderata #2 and #3. Hobbes is, of course, well known for holding that all conceptions, whether in whole or in part, have their origin in sense experience. However, Hobbes is keen to point out a limitation of relying solely on experience-based prudence; even a slight difference between past associations and a current situation can frustratingly prevent successful predictions. As he says, "[. . .] the omission of every little circumstance altering the effect, frustrateth the expectation of the most Prudent" (2012, 1052; 1651, 367). Hobbes allows experience to play a role within the two sciences of geometry and civil philosophy, but its role is limited: it can confirm the results of *scientia*, but it can never be used to disconfirm them. I argue in chapter 4 that Hobbes's reply to the individual who does not "trust" the "inference" made in *Leviathan* 13 encourages the doubter to check their experiences to receive *confirmation* of the results of the demonstration. There Hobbes encourages the reader to "consider within himselfe, when taking a journey, [that] he armes himselfe, and seeks to go well accompanied [. . .]" (Hobbes 2012, 194; 1651, 62).

Although he tolerates the use of experience for confirming the results of a scientific demonstration, he rejects any attempt to *disconfirm* results with such an appeal. For example, in *Leviathan* 20 Hobbes describes the "greatest objection" to his account of sovereign power from those who have never seen such a commonwealth: "[. . .] an argument from the Practise of men, that have not sifted to the bottom, and with exact reason weighed the causes, and nature of Commonwealths, and suffer daily those miseries, that proceed from the ignorance thereof, is invalid" (Hobbes 2012, 320; 1651, 107). Regardless of the usefulness of prudence in everyday life, in the *science* of civil philosophy Hobbes rejects it out of hand, retorting that even if "in all places of the world, men should lay the foundation of their houses on the sand, it could not be thence inferred, that so it ought to be" (2012, 320–322; 1651, 107). Unlike prudence-based practice, which he likens to "Tennis-play," Hobbes argues that the "skill [1668 Latin: *scientia*] of making, and maintaining Common-wealths" consists in "certain Rules," just like those followed in arithmetic and geometry (2012, 320–322; 1651, 107).

Beyond these approved and disapproved uses of experience in the two Hobbesian sciences, Hobbes also appeals to experience within explanations throughout other parts of his system. In particular, he appeals to experience, whether in the form of everyday experiences or natural or civil history, within natural philosophy and the parts of history that are concerned with explaining past events. This feature of Hobbes's system relates to desiderata #2 and #4. Hobbes's uses of experience in these disciplines, alongside his invocations of causal principles, whether about human nature or about geometric bodies generally, are

INTRODUCTION: THE UNITY OF HOBBES'S PHILOSOPHY   11

puzzling for the deductivist view. As mentioned already, the example of 'endeavor' is a case in point. Although 'endeavor' is introduced in natural philosophy, when used to describe the human passions, information must be added to it that is not contained within the concept 'endeavor'.

The maker's knowledge view offers an alternative account of the relationship of Hobbesian geometry to natural philosophy by arguing that mixed mathematics provided Hobbes with a model. In mixed mathematics, one may borrow causal principles from one science and use them in another science, but there need not be a deductive or reductive relationship between the two. Natural philosophy for Hobbes is mixed because an explanation may combine observations from experience or experiments with causal principles from geometry. As he puts it in 1658 in *De homine* X.5, "[. . .] physics (I mean true physics), that depends on geometry, is usually numbered among the mixed mathematics" (Hobbes 1994b, 42).

In *Examinatio* (1660), Hobbes shows continued concern for the nature of mixed mathematics when criticizing John Wallis's discussion of the pure/mixed-mathematics distinction in *Mathesis Universalis* (1657). Hobbes disagreed fundamentally with Wallis, and many other figures from the history of geometry, concerning the nature of 'quantity', but he also argued that Wallis's account of this distinction was mistaken.[7] According to Wallis, pure mathematics treats quantity absolutely, that is, quantity absolutely as "abstracted [*abstrahitur*]" from matter. In contrast, in mixed-mathematics disciplines, alongside the consideration of quantity, the particular subject in which a particular quantity exists is "connoted [*connotatur*]" (1657, 2–3). Wallis's point seems to be that in mixed-mathematics features of the subject do not enter as explicit considerations but are nevertheless taken to be actual. Hobbes ridicules Wallis's claim about connotation being different from consideration, wondering what would be called mixed at all if these are not the same and puzzling about what it could mean to *consider* the accidents of a subject but not the subject itself (OL IV.24). This conceptual worry that Hobbes introduces leads him to criticize Wallis's claim that mixed-mathematics disciplines offer demonstrations, arguing that demonstrations are solely the province of the geometer (this debate is discussed in more detail in chapter 5).

In Hobbes's *practice* of natural philosophy, his understanding of mixed mathematics implies that one may appeal to everyday experience or experiments for the demonstration that something happens (the 'that') and then borrow the cause (the 'why') from geometry. Hobbes shows that he embraces this model by appealing to experience and simultaneously by citing geometrical causal principles from Part III of *De corpore* throughout his natural philosophy in Part IV of that

---

[7] Hobbes's disagreement with Wallis concerning the object of geometry ('quantity'), which Hobbes articulates in *Examinatio* through speaker A, defines it as the determined magnitude of a body (1660, 10; OL IV.16–17). For discussion of Hobbes's rejection of the abstractionist account of 'quantity', see Jesseph (1999, 76–85).

work and also in *Dialogus Physicus*. The result of this account, which is the subject of chapter 5, is that explanations in Hobbesian natural philosophy have suppositional certainty: the natural philosopher knows that *if* the natural bodies in question behave according to the geometrically treated motion being borrowed, *then* the effects will follow of necessity. Natural philosophy—as a mixture of *cognitio* and *scientia*—thus lies epistemically between those two types of knowledge. As such, natural philosophy is not a distinct category of knowledge; instead, Hobbes understands it as a method of philosophizing that combines scientific knowledge with observations from experience.

The maker's knowledge view holds that this same pure/mixed relationship illuminates Hobbes's aim when he provides explanations—by providing the reason *why*—of historical events. Rather than understanding Hobbes's statement that, even though not on par with philosophy, civil history is "useful (no, indeed necessary) for philosophy" (Hobbes 1981, 189; OL I.9; EW I.10–11) as asserting a need for *preparation* in history by would-be political philosophers, chapter 6 argues that in explanations of the historical past *both* civil history and civil philosophy play an essential role (I call explanations of this sort Hobbesian historiography). While civil history provides the demonstration that some event has happened (the 'that'), civil philosophy provides a causal principle that can be borrowed to show the reason why (the 'why'). I argue that the causal explanations in Chapter 45 of *Leviathan* Part IV are structurally analogous to those in the natural philosophy of *De corpore* Part IV; likewise, I argue that the same relationship holds between the historiographical dialogue *Behemoth* and the natural-philosophical dialogue *Dialogus Physicus*. As with explanations in Hobbes's natural philosophy, these explanations of historical events achieve suppositional certainty by borrowing a causal principle from Hobbesian science, in this case the science of civil philosophy. While not reaching the level of certainty found in the two Hobbesian sciences, these explanations of historical events, and those in natural philosophy, are elevated above mere sense-based prudence.

The results of this account of Hobbes's two sciences and the mixed-mathematics approach reshape our understanding of Hobbes's philosophy generally and of his masterpiece the *Leviathan* in three ways. First, properly understanding Hobbes's maker's knowledge epistemology, and his corresponding view of the nature of demonstrations as synthetic constructions, exposes the deficiencies of existing accounts of the Laws of Nature. Rather than taking the demonstration of the Laws of Nature in the *Leviathan* to be a derivation in Euclidean fashion from some axiom, it shows that Hobbes saw them as part of a synthetic construction that provides readers with scientific knowledge.

Second, this account transforms our understanding of Hobbes's debate with Robert Boyle. In the light of the maker's knowledge account of Hobbes's philosophy, this debate primarily concerns the proper dependence of natural philosophy upon geometry and not the nature or value of experiment as such. Instead of

INTRODUCTION: THE UNITY OF HOBBES'S PHILOSOPHY   13

viewing Hobbes as a naïve "deductivist," Hobbes emerges as one concerned predominately with the legitimate use of geometry in natural philosophy, even alongside everyday experience and experiments. Finally, seeing mixed mathematics as Hobbes's inspiration for how disciplines fit together helps explain not only the relationship between geometry and natural philosophy but also the relationship between the science of civil philosophy and his explanations of historical events. Support for seeing this relationship as Hobbes's inspiration is found both in his descriptions of his philosophy as well as in his explanatory practice.

1.3 The Maker's Knowledge View and Developmental Worries

Scholars have wondered about Hobbes's development since at least the time of Aubrey's claim about Hobbes stumbling upon an open copy of Euclid's *Elements*. Given the extensive period over which Hobbes wrote, as well as the different geopolitical contexts in which he lived, it would be no surprise to find that he changed his mind. Indeed, it is uncontroversial that Hobbes changed his mind about *some* things, many of them minor. For example, in *Elements of Law* (hereafter *Elements*) Hobbes posited that lucid bodies like the sun move by dilation (EL I.II.8), but later in *De corpore* XXVII.2 he proposed "simple circular motion" (OL I.364; EW I.448) as responsible for their motion. This is a small change (discussed below in chapter 5).

An influential strategy defended by Quentin Skinner accommodates Hobbes's comments over the span of decades by suggesting that in *Leviathan* Hobbes fundamentally changed his mind about the nature of civil philosophy. Skinner begins by examining the Epistolary Dedicatory of *Elements* where Hobbes expresses a desire to follow "mathematical" learning because it is "free from controversies and dispute" (Hobbes 1640/1928, xvii; see Skinner 2002, 74). The new perspective Hobbes adopts, on Skinner's view, rejects the humanist conception of civil science that was grounded in eloquence and rhetoric and understands the nature of justice in terms of "right reason" alone. Skinner argues that *De cive* is similarly founded, seeing in it "a yet more confident effort to challenge and supersede the presuppositions of humanist civil science" (2002, 76). Skinner's Hobbes saw the science of civil philosophy as operating by reason alone without a need for persuasion by means of artifice or rhetorical ornamentation.

After highlighting Hobbes's self-assured descriptions of his own method and the place of reason in the science of civil philosophy in these earlier works, Skinner argues that by the time of *Leviathan* Hobbes gave up on the ability of reason to persuade readers. While not entirely abandoning the idea that civil philosophy should be demonstrative, Skinner's Hobbes now saw the proofs of reason alone as unable to persuade human beings "whose passions and ignorance lead them to repudiate even the clearest scientific proofs" (2002, 81–82). As the beginnings

14   HOBBES'S TWO SCIENCES

of this new view of civil philosophy, Skinner draws upon Hobbes's claim in *Leviathan* 10 that the sciences are "small Power" and further asserts that Hobbes includes "RHETORIQUE" as "among the most eminent faculties of the human mind" (2002, 83) in the Table of *Leviathan* 9. However, if Skinner were correct about "RHETORIQUE" gaining such eminence in *Leviathan* from its inclusion in the Table, Hobbes would have to hold that the other disciplines at the right-most part of the Table in *Leviathan* 9 were also "eminent faculties," including "MUSIQUE" and "PO-ETRY" (2012, 131). Needless to say, music and poetry are not included in any of Hobbes's explicit discussions of the nature of *scientia*, whether geometry or civil philosophy, so Skinner's appeal to the Table gives little reason to think that Hobbes came to hold rhetoric in such high esteem.

What about Hobbes's description of the sciences as "small Power"? Rather than turning back to eloquence and the art of rhetoric, the "maker's knowledge" view of civil philosophy draws upon both *De corpore* and *Leviathan* to articulate Hobbesian *scientia* as possible only if one has leisure, the proper method, and is driven by curiosity (discussed in chapter 3). This has the consequence that the scientific demonstration in the *Leviathan* was meant for select individuals. While Skinner rightly draws attention to Hobbes's claim in *Leviathan* 10 that "[t]he Sciences are small Power" while "Eloquence is Power" (Skinner 2002, 83), Skinner leaves out Hobbes's explanation for *why* the sciences are "small Power." Rather than thinking of the sciences as "small Power" *absolutely* (in contrast to eloquence), Hobbes explains that the sciences are wrongly *perceived* as lacking in power because few people understand science and "none can understand it to be, but such as in good measure have attayned it" (Hobbes 2012, 134; 1651, 42). Hobbes slightly alters this in the 1668 Latin as follows: "Science, is power [*Scientia, Potentia est*]; but a small one, because conspicuous science is rare, and therefore is not apparent except in very few people [. . .]" (Hobbes 2012, 135, trans. in fn. 10; 1668, 44). This perception of the power of science as "small" is a mistake by the "vulgar," and Hobbes provides an example of those who engage in "Arts of publique use," such as those that make instruments of war. The "vulgar" may think that such individuals are responsible for the end products like fortifications, but Hobbes holds that ultimately "the true Mother of them, be Science, name the Mathematiques" (2012, 134; 1651, 42). In other words, regardless of public perception, Hobbes held that science is *actual* power.

Hobbes's explanation for why eloquence is "Power", also not addressed by Skinner, provides additional support for the maker's knowledge view by showing that Hobbes's worries regarding its use remained even while he wrote the *Leviathan*, as Hobbes notes: "Eloquence is power; because it is *seeming Prudence*" (2012, 134; 1651, 42; emphasis added). These comments concerning the relative "Power" of science compared to eloquence show Hobbes's recognition that eloquence may sometimes be more powerful sociologically, but the "maker's knowledge" view understands the best method of persuasion as demonstrating to someone how to make something they already desire to make (in the case of civil philosophy, the maker's knowledge view articulates this desired end as the construction of PEACE). Given the many worries that Hobbes expresses regarding the limits of prudence, which I discuss in chapter 2, Hobbes's identification of eloquence as mere "seeming" prudence strongly weighs against seeing it as part of the science of civil philosophy in *Leviathan*.

A final point of contrast to Skinner's view of the accommodation of eloquence within Hobbes's civil philosophy is that it has difficulty being harmonized with Hobbes's mature understanding of demonstration in *De corpore*. Although *De corpore* was published after *Leviathan*, Hobbes worked on it before and after the latter work. While the rhetorical ends of *Leviathan* and *De corpore* differ, the "maker's knowledge" view provides the ability to see Hobbesian *scientia* as consistent between these two works. According to the maker's knowledge understanding of the science of Hobbesian civil philosophy, Hobbes's major works—*Elements, De cive*, and *Leviathan*—begin the science of civil philosophy by considering humans in their natural state and proceed by a synthetic demonstration from simples to the complex whole of PEACE. The Hobbes of *Leviathan* recognized that not everyone would be able to follow such a complex demonstration—what he saw may seem to be a "subtile deduction"—and thus allowed the Laws of Nature to be "contracted into one easie sum" (Hobbes 2012, 240; 1651, 79). Even this allowance, however, engaged the *reasoning* capacity of even the "meanest brute" and did not rely upon rhetorical artifice but instead allowed that "brute" to see that the Laws of Nature are "very reasonable." In short, the answer to Skinner's question ("How can we hope to win attention and consent, especially from those whose passions and ignorance lead them to repudiate even the clearest scientific proofs?") is that Hobbes believed that civil philosophy should meet human beings where they are in terms of their reasoning abilities, whether that amounts to giving them an "easie sum" or a complete synthetic demonstration.

2

Prudential Knowledge from Sense and the Mechanical Mind

Hobbes opens *Leviathan* with the well-known declaration that all conceptions, his frequent term for ideas, either come from sense or are derived from those that do (2012, 22; 1651, 3). All human knowledge is drawn from those conceptions. Hobbes's claim that conceptions are simply "image or fancy" is part of his strategy to do away with Scholastic faculties, like the active intellect, and explain all acts of the mind by appeal only to the imagination working with conceptions.[1] The Hobbesian mind, whose acts are made intelligible by appeal only to bodies in motion, is a mechanical mind. This reduction is evident in *Leviathan* but also in works that precede it, such as *Anti-White* IV.1 (c. 1642–1643) where Hobbes reduces talk of being "in the intellect" simply to being "an imaginary thing" (Hobbes 1973, 126; 1976, 52).

In understanding conceptions as images, Hobbes makes clear that "the object is one thing, the image or fancy is another" (2012, 24; 1651, 3). Conceptions may resemble external objects or they may not. This gap between conceptions and external objects may seem to place Hobbesian epistemology in a worrisome position. Some have seen Hobbes as moved in the direction of *epistemological* idealism, despite his materialist metaphysics (Guyer and Horstman 2015), or as anticipating features of Kant's transcendental idealism, especially with regard to the account of 'space' (Herbert 1987).[2] Others have taken Hobbes to be worried about external world skepticism like Descartes's meditator was. For example, Richard Tuck claims Hobbes's epistemology was "partly designed to meet Descartes' hyperbolic doubt" (Tuck 1988, 37–41).[3]

---

[1] Hobbes's philosophical psychology developed in early works such as *Elements of Law* (1640) and *Anti-White* (1642–1643) informed his well-known objections to Descartes's *Meditations on First Philosophy* (for discussion, see Adams 2014a).

[2] For criticism of Herbert's Kantian interpretation of Hobbesian 'space', see Jesseph (2016, 136) and Leijenhorst (2002, 107–108).

[3] Tuck's interpretive claim that Hobbes sought to rebut Descartes's worries, what Tuck sees in Hobbes's ability to "take on board the hyperbolical doubt without having to turn to the proof of God's existence to do it" (1988, 37), overstates the evidence provided in support of it. After recognizing that "[i]t may be that [Hobbes] developed his original idea independently of both Gassendi and Descartes" (1988, 37), Tuck appeals to texts (see 1988, 37–41) that relate not to Cartesian hyperbolic doubt but rather to Hobbes's worries about securing knowledge of existing bodies' *actual* motions rather than mere *possible* motions. For example, as support for the claim that Hobbes thought it was impossible to "secure knowledge of the external world," Tuck (1988, 37) cites a letter from Hobbes to William Cavendish written 29 July/8 August 1636: "In things that are not demonstrable, which kind is ye

---

Hobbes's Two Sciences. Marcus P. Adams, Oxford University Press. © Marcus P. Adams 2025.
DOI: 10.1093/9780198924715.003.0002

However, Hobbes is unconcerned with global worries regarding the existence of an external world. Hobbes makes this clear by choosing to ground "first philosophy" in *De corpore* VII with a thought experiment imagining a man who remains after the annihilation of the entire world. Implicit in the thought experiment is the assumption that the world that caused the conceptions remaining in the man's mind did exist at one time (see Sorell 1995, 92).[4] Everyday human knowers are in the same position as that imagined individual; in either circumstance, a knower must reason on the basis of conceptions caused by external objects. As Hobbes puts it, whether reasoning about existent or nonexistent things, "we compute nothing other than our phantasms" (OL I.82; EW I.92).[5] Similarly, Hobbes's explanations of sensation, such as in *De corpore* XXV, suppose that motions from external bodies and resistance to those motions from perceivers' bodies are the cause of sense.[6]

Such a starting point for philosophy would clearly be unsatisfactory from a Cartesian perspective. The issue at stake for Hobbesian epistemology is that the actual causes of natural phenomena are underdetermined by the observed effects. Hobbes never doubts whether the external world exists but instead is concerned about what humans can know about that world given its existence.[7] Hobbes's lack of concern for Cartesian skeptical worries is furthermore clear from his treatment of dreams in *Leviathan* 2:

greatest part of Naturall Philosophy, as depending vpon the motion of bodies so subtile as they are inuisible, such as are ayre and spirits, the most that can be atteyned vnto is to haue such as opinions, as no certayne experience can confute, and from which can be deduced by lawfull argumentation, no absurdity [...]" (Hobbes 1994a, 33). Instead of conveying skepticism about knowledge of the external world, *pace* Tuck Hobbes expresses only that, because the motions of *some* bodies that cause natural phenomena are so small as to avoid observation, merely opinions and not causal knowledge may be reached.

[4] Leijenhorst suggests that Hobbes's claim that perception and perceptual change are constituted by motion grounds the belief that there is a mind-independent reality such that "concepts are linked to the world [...] by their being motion in our bodies, and not by their being representations" (2002, 88–89). However, Leijenhorst worries that this may be insufficient to prevent Hobbes from falling into solipsism since 'causality' is a principle of reason. This is indeed a worry for Hobbesian epistemology, especially since one could never be absolutely sure that the motions constitutive of perception were not from some deceptive cause. Nevertheless, although solipsism may be a *conceptual* possibility, Hobbes ultimately shows little concern for worries like this.

[5] Nuchelmans (1983, 131) connects Hobbes's understanding of computation to Ramus. For discussion of Ramus's account of syllogizing and ratiocinating, and the lineage of those views in financial bookkeeping, see Nuchelmans (1980, 168–169).

[6] Chapter 5 discusses this explanation of sense as well as other suppositions Hobbes makes throughout Part IV, such as the six suppositions in *De corpore* XXVI.5.

[7] Agreeing that Hobbes is unconcerned with Cartesian worries, McNeilly goes further by claiming that Hobbes simply makes "dogmatic assumptions" and that "Chapters I and II of *Leviathan* ['OF SENSE' and 'OF IMAGINATION'] represent [...] an area in which [Hobbes] had no philosophy at all" (1968, 31). However, Hobbes's deep concern with issues in philosophy of perception is abundantly clear from his abiding interests in optics and perception across numerous works. For discussion of these aspects of his thought, see Adams (2014b; 2016a; 2016b), Giudice (2016), Malet (2001), Médina (2016), Prins (1987; 1993; 1996), and Shapiro (1973).

18   HOBBES'S TWO SCIENCES

> And because waking I often observe the absurdity of Dreames, but never dream of the absurdities of my waking Thoughts; I am well satisfied, that being awake, I know I dreame not; though when I dreame, I think my selfe awake. (2012, 30; 1651, 6)

In the Third Set of Objections, Hobbes dismisses Descartes's worries about distinguishing dreams from waking veridical sensations out of hand as simply rehashing "ancient material." Hobbes admits that "if we follow our senses, without exercising our reason in any way, we shall be justified in doubting whether anything exists" (CSM II.121; AT VII.171), but Hobbes's implication is that we are not simply stuck in instantaneous moments with worries like "Am I dreaming *now*?" and a few seconds later "What about *now*?" We can simply reason, as in *Leviathan* 2, that when I am awake, I know am not dreaming (since I "often observe the absurdities of my dreams"), but that when I am asleep, I might take myself falsely to be awake. This bit of reasoning shows Hobbes's view that failing to engage the skeptic does not threaten the possibility of knowledge. From the Cartesian perspective this fails to engage with the skeptic, but Hobbes's approach is justified given his aims. Since he understood philosophy as akin to corn and wine, existing naturally but lacking in cultivation, he believed that the goal of philosophy was only to help improve "natural reason" (OL I.1; EW I.1–2). Such an approach is clearly not available to Descartes's meditator.

This chapter begins a two-chapter account of Hobbes's epistemology by focusing first upon what Hobbes frequently identifies with the Latin term *cognitio*: human and non-human knowledge founded in conceptions received in sense. I address (1) the source of Hobbesian *cognitio* from sense through prudence as particular conceptions among which a knower judges "similitudes" (an inference from many to all or most), (2) Hobbes's explanation for this behavior as seeking to bring about antecedents of desired consequents, and (3) the role that language plays in allowing human knowers to mark conceptions with names, form contingent propositions and syllogisms using those propositions, and raise conceptions in others by using names as public signs. In chapter 3, I argue that Hobbesian epistemology distinguishes *cognitio* from *scientia*—certain, demonstrable knowledge—by proposing a different source, role for language, and explanation for the behavior that brings it about.

2.1 Conceptions from Sense and *Cognitio*

Hobbes's explanation of sense and mental acts such as imagination, experience, and memory is consistent among early works, such as the *Elements of Law* (1640; hereafter *Elements*) and *Anti-White* (c. 1642–1643), and later works, such as *Leviathan* (1651) and *De corpore* (1655). His account is mechanical insofar as it appeals only to bodies in motion and rejects the explanatory efficacy of distinct

PRUDENTIAL KNOWLEDGE FROM SENSE AND THE MECHANICAL MIND   19

faculties and of visible *species* (EL II.4; Hobbes 2012, 24; 1661, 4).[8] All mental acts, Hobbes claims, can be understood without appeal to distinct faculties, such as the intellect or the will, and simply in terms of motion. Hobbes emphatically calls these *acts* rather than faculties when, for example, he expounds upon the nature of the will in *Leviathan* 6 (2012, 92; 1651, 28). Rather than relying upon the faculty of the intellect and its judgments, compared to the will, as Descartes's meditator does in Meditation IV (CSM II.43; AT VII.62), Hobbesian sense can correct *itself* with repeated and varied experiences (e.g., EL II.10; EL II.6; also 2012, 24; 1651, 3–4).

Conceptions received in sense are the foundation of *cognitio* throughout Hobbes's corpus. Beginning in early notes for *De corpore* (c. 1637), Hobbes likens the mind to a mirror: "The mind of man is a mirror capable of receiving the representation and image of all the world" (ms. 5297; Hobbes 1973, 449).[9] Sense, Hobbes argues in *Elements*, produces "certain images or conceptions of the things without us" (EL I.8) and these "proceed from the actions of the thing itself, whereof it is the conception" (EL I.2). In *Anti-White* XXX.3, Hobbes argues that conceptions are produced because bodies press upon sense organs either immediately in taste and touch or mediately in sight, hearing, and smell (1973, 349; 1976, 364–365). Although there is broad consistency between Hobbes's earliest and later philosophical psychology, the beginning chapters of *Elements* provide material on the nature of conceptions that is not repeated in *Leviathan*, and they use the term 'conception', which is not a central term used in the *Leviathan* account.[10] Given the difference of material and apparent change of terminology, it may be tempting to regard Hobbes as changing his mind, but I take this additional material in *Elements*, where Hobbes frequently uses the term 'conception', to be consistent with Hobbes's later claims and illuminative of his understanding of *cognitio*. Thus, I briefly focus on this aspect of *Elements* before using later works alongside it.

---

[8] For discussion of Hobbes's critique of faculties, see Leijenhorst (2002, 89ff). In contrast, Chadwick (2020) argues that in the early *Elements of Law* Hobbes uses 'mind' and 'soul' to contrast natural faculties with those related to soteriological matters. Hobbes elsewhere appeals to a "natural light" (*lumen naturale*) (e.g., OL IV.95; OL IV.395; OL IV.446) that grasps mathematical definitions, but these references can be understood as referring to certain principles that are graspable without analysis. I thank Douglas Jesseph for drawing my attention to these texts.

[9] Malcolm (1996, fns. 49 and 70) dates ms. 5297 to 1642–1643, but Pacchi (1965) and Rossi (1942) have suggested earlier dates. Even if these notes were written as late as 1642–1643, Hobbes likely began working on his philosophical psychology as early as 1637: in a letter written in January of that year, Digby refers to Hobbes's work on "conceptions" in his "Logike" (Hobbes 1994a, 42), and in a letter written in September of that same year, he begs Hobbes to send him any part of his "Logike" as soon as it is completed (Hobbes 1994a, 50). Malcolm speculates that "Logike" refers to ms. 5297 (Hobbes 1994a, 43 fn. 2).

[10] This is not meant to imply that Hobbes never uses the term 'conception' in *Leviathan*; indeed, he uses it within the definition of 'sense' (2012, 22; 1651, 3). My point is that Hobbes uses 'conception' more frequently in *Elements* than when discussing similar topics in *Leviathan*, the latter of which often uses 'thoughts' (often translated in the 1668 Latin with *cogitatio*), and 'thought' is not among the "words equivalent" to 'conception' in *Elements* (I.1.8). Unlike in *Elements*, where talk of 'conception' dominates the early chapters, Hobbes waits until *Leviathan* 45 to describe the nature of an image or phantasm, and he offers that account only in midst of a discussion of idolatry (2012, 1030–1032; 1651, 358).

20   HOBBES'S TWO SCIENCES

Both *Leviathan* 1 and *Elements* I.2 include discussion of 'sense', but in *Elements* a chapter titled "The general division of man's natural faculties" precedes the account of sense. Hobbes outlines how "contained in the definition of man" are the "natural faculties and powers" of humans (EL I.1.4). These divide into the faculties of body and mind (EL I.1.5), and he further divides "powers of the mind" into "cognitive or imaginative or conceptive; and motive" (EL I.1.7).

Concerning this "power cognitive," Hobbes asserts that even if the "rest of the world were annihilated" conceptions as images would remain, holding that "every man by his own experience knowing that the absence or destruction of things once imagined, doth not cause the absence or destruction of the imagination itself" (EL I.1.8).[11] As later in *De corpore* I.3 (OL I.4; EW I.4–5; Hobbes 1981, 179) where Hobbes uses 'conception' and 'idea' as synonyms, in *Elements* he states that "cognition, imagination, ideas, notice, conception, or knowledge" all refer to "[t]his imagery and representations of the qualities of things without us" (EL I.1.8). Whatever it is in human perceivers that makes them capable of such representation just is the "cognitive power" (EL I.1.8). Hobbes's use of 'faculty' may seem to contradict my earlier claim that he rejects talk of distinct faculties, but I suggest that it implies only that the mind is able to perform various acts and not that there is some essential difference in kind between the mind's acts.[12] Although sense may seem to differ in *kind* from the mind's other acts because an object is present for sense but absent for all others, considered simply as motion, the act of sense and all further acts of the mind differ from one another only in *degree*. All mental acts differ from one another in terms of how much the motions constituting them have decayed from the original motions of sense.

Focusing on Hobbes's frequent use of the term 'conception' in *Elements* exposes his early, detailed account of the mind there. Simply noticing that material and phrasing from the early chapters of *Elements* differs from the *Leviathan* would be unremarkable since there are many other differences between those works. Hobbes shows the importance of 'conception' for his philosophical psychology with an extended discussion in *Elements* I.2, and 'conception' recurs repeatedly in *Elements* I.3–8. Throughout *De corpore* Hobbes likewise frequently refers to conceptions, most prominently in his discussion of the "simplest conceptions" that are "universals" in *De corpore* VI.6, so examining Hobbes's preoccupation with 'conception' in *Elements* provides a link to *De corpore*.

---

[11] This is an early version of the annihilation of the world thought experiment that Hobbes uses to ground first philosophy in *De corpore* VII (discussed in chapter 4).

[12] Hobbes uses 'faculty' elsewhere simply to describe an ability, such as the "faculty of speaking powerfully" (EL II.9.14) and the "faculty of imposing names" (EL I.9.18). Similarly, in *Leviathan* Hobbes describes goal-orientation discourse of the mind as "*Seeking*, or the faculty of Invention" (2012, 42; 1651, 9) and highlights humans' "excel[ling] all other Animals in [the] faculty" of being "apt to enquire into the consequences of [something conceived], and what effects he could do with it" (2012, 68; 1651, 20).

Elements I.2 begins by describing the "difference, their causes, and the manner of their production [. . .]" of conceptions (EW I.1.1). According to both *Elements* and *Leviathan*, sense is the "originall" of all mental thoughts or conceptions (EL I.2.2; 2012, 22; 1651, 3) and both distinguish what each sense organ receives, such as "light" and "sound" (the "qualities called *Sensibile*"; see 2012, 22–24; 1651, 3). However, though *Leviathan* identifies these as "seeming, or *fancy*" that "consisteth" in these qualities from various senses, it does not explicitly identify each of these productions as a distinct conception. In contrast, the account in *Elements* states that each sense organ yields a separate, particular conception of an external object: "By our several organs we have several conceptions of several qualities in the objects" (EL I.2.3).[13] *Elements* continues by calling each representation produced in the sense organs a conception, so that of the same object by sight "we have a conception or image composed of colour or figure," by hearing "we have a conception called sound," and so on (EL I.2.3). Perceiving a bouncing red ball, for example, would yield multiple conceptions: by sight the conception of an apparently red surface, by hearing the conception of the sound of a bounce, and by touch the conception of resistance and texture. These conceptions would multiply as time progressed, so by sight the conception of an apparently red surface occurs at time T_1 but, say, at T_2 the angle of the light illuminating the ball changes so a distinct conception of an apparently reddish-orange surface results. Hobbes does not say how it happens, but he assumes that sense, without a separate faculty of "common sense," joins together conceptions from each sense so that experiences of external objects are taken to be of unified bodies. I will refer to these collections of the accidents of bodies[14] we perceive as unified, conglomerate conceptions.[15]

---

[13] The explanation of sense in *De corpore* XXV.10 likewise distinguishes what each sense organ receives, using the synonymous term 'phantasm' rather than 'conception'. Hobbes notes that the "proper phantasm [*phantasma*] from sight is light [*lumen*]" (OL I.329; EW I.404) and the "phantasm from hearing, sound; from smell, odor; from taste, savor: and from touch are hardness and softness, heat and cold [. . .]" (OL I.330; EW I.405). The terminology of proper *phantasms* rejects the Scholastic understanding of the proper *objects* of the senses, such as that the proper object of vision is light and color (Leijenhorst 2002, 87). Against that view, Hobbes claims bodies themselves are proper objects of the senses.

[14] Hobbes says in *De corpore* VIII.2 that "what an accident is cannot so easily be explained by any definition, as by examples" (OL I.91; EW I.102), so he asks the reader to imagine a body as filling a part of space such that it is "coextended with it." He argues that this coextension is distinct from that imagined body itself and must be an accident. Likewise, motion must be an accident because when imagining that body being moved, the "removing [of the body] is not the removed body" itself (OL I.91; EW I.102). Hobbes uses these considerations to criticize those who claim that "an accident is something, namely, some part of a natural thing" (OL I.91; EW I.103). Instead of understanding accidents as parts of bodies, Hobbes argues that an accident is "the manner by which any body is conceived," which he equates to understanding accidents as faculties of bodies to cause conceptions in perceivers (OL I.91; EW I.103). Were accidents actually *in* bodies as parts, then they would have an independent existence, and this would require that "an accident would be a body also" (OL I.92; EW I.104). Similarly, it would be "improper to say, an accident is moved"; rather than saying that "figure is an accident of body carried away," we should say that "a body carries away its figure" (OL I.104; EW I.117).

[15] Hobbes agrees with Scholastic views that understand each sense as receiving, in Hobbes's terms, "qualities called *Sensibile*" that are proper to it, but he disagrees that his philosophical psychology

22   HOBBES'S TWO SCIENCES

Later in *De corpore* I.3, Hobbes similarly describes how the mind automatically puts together and takes apart conceptions without the use of words, providing examples of "how we are accustomed to add or subtract mentally, reasoning in silent thought without words" (OL I.3; EW I.3; Hobbes 1981, 177). The example Hobbes provides is a case of seeing something from afar, and it draws upon the senses of sight and sound. The point Hobbes makes there is the same as in *Elements* I.2.3: the human mind automatically joins together conceptions from disparate sense organs into a unified, conglomerate conception. He argues that when seeing something at a distance "obscurely" the mind receives the conception BODY.[16] Upon observing that thing move, the mind receives ANIMATE, and upon hearing that mobile thing speak and observing its behavior up close, the mind receives RATIONAL. Having received these conceptions from sight and hearing, Hobbes claims that when "one conceives the whole thing seen as one, fully and distinctly, the idea [*idea*] is composed from the preceding ones" (OL I.3–4; EW I.4; Hobbes 1981, 177–179). Although Hobbes connects this automatic composition of particular conceptions from separate senses into unified, conglomerate conceptions to the eventual names that may later be applied (e.g., with "[...] no words imposed, still he has the same idea of that thing, and because of that he calls the thing 'body' when words are imposed"; OL I.3; EW I.4; Hobbes 1981, 177), he is clear that this example shows "how the internal reasoning of the mind works without words [*sine vocibus*]" (OL I.4; EW I.5; Hobbes 1981, 179). Were readers just to attend to the explanation of sense in *Leviathan*, they may hold that in sense a single conception—simply "the image or fancy" (2012, 24; 1651, 4)—is formed, such as the single conception RED-BOUNCING-BALL. The account Hobbes offers in *Elements* and *De corpore*, which distinguishes each sense as receiving its own conceptions, however, suggests a more complex picture of how conceptions are received in sense: in an act of sense the mind joins together conceptions to produce something unified (the conception RED + BOUNCING

---

requires a separate faculty of common sense to join together conceptions from each sense and make judgments. See Leijenhorst for discussion of this background (2002, 82–83). Indeed, Hobbes includes the memory and judgment of phantasms in the definition of 'sense' in *De corpore* XXV.5 (OL I.320; EW I.393). Barnouw takes Hobbes's assertion that this definition of 'sense' reflects how sense is commonly understood as "surprising" (1980a, 122). Hobbes's ridicule in *Leviathan* 2 of Scholastic appeals to a common sense as an example of "many words making nothing understood" (2012, 36; 1651, 8) does make his assertion that he is providing the typical view of sense seem disingenuous, but Leijenhorst criticizes Barnouw's surprise and argues that Hobbes's claim that sense judges agrees with many Scholastic accounts but disagrees that a separate faculty is needed for it (2002, 82 fn. 124). Heinrichs highlights Hobbes's assumption that the process of individual conceptions from each sense combining together must happen rapidly to give subjects the experience of "a complete and compounded picture of the object" (1973, 60)

[16] I use upper-case letters when referring to conceptions and lower-case letters in single quotation marks when mentioning words.

SOUND + RUBBER TEXTURE + and so on), and it does so instantaneously without a perceiver's awareness.

Following the account of the conceptions produced by each sense, *Elements* offers "four propositions concerning the nature of conceptions" (EL I.2.4–9). These propositions are meant to show that these conceptions that we have from various senses are not in objects (proposition 1), nothing outside of us (proposition 2), nothing but "apparitions" of motions from objects (proposition 3), and that all conceptions are not in objects but in perceiving subjects (proposition 4). Although this material is not directly included in *Leviathan*, the account there agrees with these propositions as Hobbes declares that "[…] the object is one thing, the image or fancy is another" (2012, 24; 1651, 3).

The final aspect of Hobbes's view of the conceptions received in sense is that each of them encodes that it was caused by something outside of a perceiver. Hobbes posits that this encoding is due to an outward rebound of motion from a perceiver that resists inward motions from objects:

> Now the interior coat of the eye is nothing else but a piece of the optic nerve, and therefore the motion is still continued thereby into the brain, and by resistance or reaction of the brain, is also a rebound in the optic nerve again, which we not conceiving as motion or rebound from within, think it is without, and call it light […] (EL I.2.8)

These motions "reboundeth back into the nerves outward, and thence it becometh an apparition without" (EL I.2.9).[17] Likewise, in *Leviathan* 1 Hobbes appeals to "resistance, or counter-pressure" that causes perceivers to take conceptions of external objects as "seemeth to be some matter without" (2012, 22; 1651, 3; see also *Anti-White* XXX.3, Hobbes 1973, 349–350). Hobbes's most detailed explanation of sense in *De corpore* XXV.1–2, discussed in chapter 5, also includes resistance in its definition of 'sense' (OL I.319; EW I.391).

Hobbes's account of the conceptions received in sense discussed so far has four key features. First, sense is the source of all conceptions—all conceptions are either directly from sense or derived from those that are. Second, conceptions are constituted by motions that each sense organ receives, so in sensing a single external object a perceiver will possess distinct conceptions from the senses of sight, hearing, touch, taste, and smell. Third, sense automatically joins together these conceptions from each organ into unified, conglomerate conceptions. Finally,

---

[17] Unlike in *De corpore* XXIX.1, where the heart is responsible for the outward motion in *both* conceptions and passions (OL I.396; EW I.486), in *Elements* the brain alone is responsible for the outward motion that forms the conception that a body is outside of us. In *Elements*, the heart is involved in the outward motions of the passions and "conception is nothing but motion within the head" (EL I.8.1). Pettit neglects this aspect of the Hobbesian mind (2008, 13–16); without it, Hobbes cannot explain perception of *external* objects.

24   HOBBES'S TWO SCIENCES

Hobbes posits a resistance to the inward motion from external objects that causes motion outward, which accounts for why human perceivers take external bodies to be the cause.

2.2 Motions Remaining in the Mechanical Mind

Hobbes explains all other mental acts beyond sense and the conceptions received in it by appeal to motion. These additional mental acts are imagination, experience, memory, and prudence. *Leviathan* explains imagination as occurring when the motions of sense decay after an external object is no longer present or a perceiver's eyes are shut (Hobbes 2012, 26; 1651, 5; cf. EL I.II.2).[18] The Latin *Leviathan* expands on the definition of imagination as "decaying sense" by adding 'phantasm', one of Hobbes's synonyms for 'conception': *sive Phantasmata dilutum & evanidum* ("or a diluted or evanescent Phantasm") (Hobbes 2012, 27; 1668, 5). Memory and imagination are both motions remaining after sense, and they differ only in how a knower considers those images; they are "but one thing, which for divers considerations hath divers names." The name 'memory' is applied when motions from sense are *viewed* as "fading, old, and past" (Hobbes 2012, 28; 1651, 5). The earlier *Anti-White* account makes this same distinction by referring to how we *consider* the same conception: "[. . .] when we are not considering [*considerare*] the past, we call the motion an imagination, but every time we do consider [*considerare*] the past we call the same motion memory" (*Anti-White* XXX.5, Hobbes 1973, 351; 1976, 366–367). Hobbes reiterates this claim in *De corpore* XXV.8 (OL I.324; EW I.398).

The definition of "PHANTASY or IMAGINATION" in *Elements* displays the continued relevance of 'conception' in the earlier account: "imagination being (to define it) conception remaining, and by little and little decaying from and after the act of sense" (EL I.3.1). Other mental acts are similarly explained by the absence of an external object and a weakening of motions, such as dreams (EL III.3), fictions of the mind (EL III.4), and memory (EL III.6–7). In *Leviathan* 2, Hobbes justifies this continuum of motions remaining in the mind after sense by appeal to an *a priori* principle about motion, which is not present in *Elements*: "When a body is once in motion, it moveth (unless something els hinder it) eternally; and whatsoever hindereth it, cannot in an instant, but in time, and by degrees quite extinguish it" (2012, 26; 1651, 4).[19] Following his invocation of this principle, Hobbes

---

[18] I focus only on the role of imagination in the production of experience, memory, and prudence and leave to the side Hobbes's worries about the potential harm the imagination makes possible, such as the role it plays in religious superstition (e.g., *Leviathan* 2, 2012, 34–36; 1651, 7–8; also *De corpore* XXV.9, EW I.402). For discussion of dreams and prophecy, see Hoekstra (2004, 138–140), and for discussion of the imagination's potential for causing political instability, see Schwartz (2020).

[19] This same principle figures throughout much of Hobbes's philosophy. Chapter 5 discusses its use in natural philosophy.

PRUDENTIAL KNOWLEDGE FROM SENSE AND THE MECHANICAL MIND   25

analogizes the continuum of motions in the mind to waves made by wind in a body of water; just like ripples in water may continue long after the wind has died down, so too the motions caused by sense experiences continue and fade little by little (see also EL I.3.1; also *Anti-White* XXX.4, Hobbes 1973, 350).

The decay of these motions explains why imagination and memory lack content originally present in sense. Decay causes the clarity and distinctness of the details in sense to fade. Hobbes provides an example of a conception of a city to illustrate this.[20] When the conception is produced as the city is sensed it is "clear" such that the conception "represented all of the parts distinctly" (EL I.3.7). A person sensing a city can "distinguish particular houses, and parts of houses," but upon departing their conception becomes obscure and is only "a mass of building" (EL I.3.7). The account of "remembrance" in *Elements* similarly emphasizes the role of conceptions (EL I.3.6–7). There Hobbes argues that remembrance, which "may be accounted a sixth sense," occurs "when the conception of the same thing cometh again, [and] we take notice that it is again" (EL I.3.6). This "internal" sense produces a conception, or "notice," as Hobbes says, "we take notice that it is again" (recall that in *Elements* Hobbes uses 'notice' as a synonym for 'conception'; cf. EL I.1.8). In other words, in the mental act of remembrance a knower forms an additional conception (a notice), namely, that they have previously considered some conception, either in sense or in imagining. The account of memory in *Leviathan* 2 uses a similar example of perceiving a city, likening memory to seeing some object from afar such that it "appears dimme, and without distinction of smaller parts" (2012, 28; 1651, 5). In contrast, when we walk the streets of a city, we observe its details in fine grain as we move through its parts. Later in the day those details fade as we try to imagine the things we sensed, and "after a great distance of time" we lose these details to an even greater extent and "wee lose [. . .] many particular Streets" (2012, 28; 1651, 5).[21] However, the *Leviathan* depiction of memory lacks reference to the additional conception (notice) that Hobbes says in *Elements* I.3.6 is formed when the mind engages in that mental act.

Adding up memory upon memory—or what Hobbes calls "[m]uch memory, or memory of many things" (Hobbes 2012, 28; 1651, 5)—results in experience. Importantly, the account of sense and imagination that Hobbes provides in *Leviathan* 1 prior to considering memory is meant to show the *genesis* of human

---

[20] In a letter from 21/31 October 1634, Hobbes explains the phenomenon of remembering a friend's face better than one's own by appealing to motions of varying amounts of "force" (Hobbes 1994a, 22).

[21] *Leviathan* might seem to make a distinction not present in *Elements*—a distinction between memory and remembrance. In contrast to Hobbes's placement of the discussion of remembrance within the *Elements* chapter on imagination (EL I.3), he discusses remembrance not in *Leviathan* 2 with memory but in *Leviathan* 3 as a mental act related to trains of "regulated thoughts" (2012, 42; 1651, 10). However, this is merely a terminological difference, for in *Elements* this same mental act is called "reminiscence" (EL I.4.5), which is aligned with Hobbes's claim in *Leviathan* 3 that remembrance is what "the Latines call [. . .] *Reminiscentia*, as it were a *Reconning* of our former actions" (2012, 42; 1651, 10).

26   HOBBES'S TWO SCIENCES

conceptions, but sense for humans in the actual world will always involve experience and the comparison of conceptions being caused by objects present with those lingering from the past. In other words, sense in the actual world by those with a ready stock of conceptions is inherently comparative and includes judgments among conceptions (Abizadeh 2017; Barnouw 1980a, 122–123). After providing the definition of sense in *De corpore* XXV.5, Hobbes recognizes this: "[...] by sense, we commonly understand the judgment we make of objects by their phantasms; namely, by comparing and distinguishing those phantasms" (OL I.320; EW I.393).

The accumulation of experiences provides a greater set of conceptions with which to make comparisons, enabling the development of prudence. Although humans are unable to have "a conception of the future," since the future has not yet happened, prudence allows us "of our conceptions of the past, [to] make a future" (EL IV.7). Hobbes's understanding of conceptions is that they are caused, whether mediately or immediately, by sense, but here he means that humans can create new complex conceptions of a *possible* future. The only cause of a conception of the actual future will be future sense itself, but Hobbes allows for current conceptions to serve as a cause of a complex conception of some possible future, much like the complex conception someone could create of the un-experienced BIG + ROCK + CANDY + MOUNTAIN. This possible future, caused by considering the aggregate of current conceptions in this way, could turn out to be false. Hobbes's use of 'decay' when describing the motions begun in sense and resulting in memory, experience, and prudence may seem to imply a negative valence, but we need not see it as such. Indeed, part of the *usefulness* of memory is that the complex details—potentially overwhelming—from the moment of sense disappear and only the salient antecedents needed for prudence remain (discussed more in subsection 2.4.1).[22]

An illustration will help clarify the way in which Hobbes thinks prudence develops. Imagine that I hear a howling sound outside at night. This may result in my making "a conjecture of things to come" (EL IV.7) such as the following: a coyote is about to eat the chickens in the chicken coop. If the sound I hear sounds sufficiently coyote-like, I will decide to check the security of the chicken coop. My conjecture is only as good as my prior experiences: those of nocturnal animal noises, my comparison between the current sound and those of the past, and the linkages related to what happened to the chickens in the coop following my hearing of past sounds from coyotes. In *Leviathan* 3, Hobbes explains these linkages as "Transitions from one imagination to another [. . .]" from past sense experiences (2012, 38; 1651, 8). The likelihood that my conjecture that the conglomerate conception "HENS + EATEN" will soon arise in sense depends upon the quantity and quality of past experiences, what Hobbes describes in *Leviathan*

---

[22] Pettit neglects this important role played by passive decay in the transition of conceptions from sense to memory, instead claiming "attention" in the mind organizes its conceptions (2008, 15).

PRUDENTIAL KNOWLEDGE FROM SENSE AND THE MECHANICAL MIND   27

3 as "*Foresight*, and *Prudence*, or *Providence*; and sometimes *Wisdome*" (2012, 42; 1651, 10).[23] Likewise, he identifies prudence in *De corpore* I.2 as "nothing other than the expectation of things similar to those which we have already experienced" (Hobbes 1981, 177; OL I.3; EW I.3).

Hobbes's well-known criticisms of the absurdities made possible by lack of care in the use of language have been the subject of scholarly discussion,[24] but Hobbes also holds, *pace* Raylor (2018, 17), that error occurs at the level of conceptions, entirely apart from the employment of language. For example, he discusses how "error of mind without use of words" occurs in *De corpore* V.1 (Hobbes 1981, 271–272; OL I.51; EW I.55–57), something also briefly mentioned in *Leviathan* 4 (2012, 54; 1651, 15). Hobbes's more detailed account in *De corpore* outlines the following types of "error in sense or thought [*sensu et cogitatione*]" and provides examples of each: "when we imagine something other than the present imagination, or when we represent as past the things which have not occurred, or as future things which will not follow" (Hobbes 1981, 271; OL I.50; EW I.56).[25] I focus only on the final type of error since Hobbes's example is a case of prudence: "[. . .] from the sight of swords that there has been or will be a fight, because it usually happens in this way for the most part" (Hobbes 1981, 271; OL I.50; EW I.56). This example of an error without the use of words transpires when one moves from a conception caused by the sight of a group of swords (SWORDS) to the expectation that another conception will soon appear in one's mind, i.e., that of a fight that will be or has been (FIGHT). Considered simply as images, Hobbes notes, these objects of the mind cannot be "false since they truly are what they are" (Hobbes 1981, 271; OL I.50; EW I.56–57); truth and falsity are applied only to propositions (see *De corpore* III.7, OL I.31; EW I.35). Error can thus arise when humans and other animals, indeed, "all things endowed with senses," make conjectures on the basis of images.[26]

One might wonder what kinds of mental attitudes an animal without language can have. Hobbes holds that all things with sense have conceptions (images) and from those conceptions (and trains of them) naturally conjecture the presence of

---

[23] I use upper-case letters to emphasize that conjectures are made on the basis of imaginations, i.e., conceptions decayed from sense, and '+' to emphasize that these are conglomerate conceptions. I discuss how such conceptions relate to language in section 2.4 below.

[24] See Duncan (2016) for recent treatment.

[25] In *System of Logic* I.V.3, J.S. Mill claims that such error would be propositional: "a person who has not the use of language at all *may form propositions mentally*, and that they may be untrue, that is, he may believe as matters of fact what are not really so. This last admission cannot be made in stronger terms than it is by Hobbes himself; though he will not allow such erroneous belief to be called falsity, but only error" (1974, 96; emphasis added). Hobbes clearly would deny that, in his terms, a proposition is formed in such cases since his goal is to explain how error *apart from words* is possible, and Hobbesian propositions employ names (OL I.27; EW I.30).

[26] Abizadeh (2017, 23–25) discusses the other two types of error and provides conceiving of a two-dimensional drawing as three-dimensional via memory, which Hobbes describes in from *A Minute or First Draught of the Optiques* (1983/1646), as an example of the first type of error.

28   HOBBES'S TWO SCIENCES

the next conception that is likely to come. In *Leviathan* 7, Hobbes reserves the mental attitude of "belief" for those using language when they find themselves in social situations where a "discourse" does not begin from definitions but where the hearer trusts someone else's testimony. In such a case, Hobbes notes "[. . .] in Beleefe are two opinions; one of the saying of the man; the other of his vertue" (2012, 100; 1651, 31). These conjectures based on images by non-linguistic animals will simply be stimulus-response reactions that guide behaviors. An error for such animals is as follows: based on some sensory input, incorrectly forming an image of something conjectured to follow and then acting upon that conjecture, i.e., taking "as future things which will not follow" (Hobbes 1981, 271; OL I.50; EW I.56).

The account in *Leviathan* reflects the uncertainty related to conjectures made from prudence. Hobbes reports that conjectures can be "very fallacious," attributing this potential for error to the "difficulty of observing all circumstances" (Hobbes 2012, 42; 1651, 10). His example of such a conjecture is when someone "foresees what wil become of a Criminal." He seems to mean that if someone saw *all* of the linkages between *every* criminal and the punishment each of those criminals had received (an induction by simple enumeration), then that person would have near certainty about conjectures like future occurrence of $CRIMINAL_a$ + GALLOWS on the basis of the conception of $CRIMINAL_a$ + $CRIME_b$.[27] Were humans able to observe "all circumstances" there would be no risk in making this conjecture. However, since humans have different experiences from one another, Hobbes states that prudence comes in degrees: "[. . .] by how much one man has more experiences of things past, than another; by so much also he is more Prudent" (2012, 42; 1651, 10). The account of prudence in *Elements* agrees, explaining "old men [as] more prudent" because they have had more experiences (EL I.4.10). Even the *most* prudent individual will be limited to conjecture since "[e]xperience concludeth nothing universally" (EL I.4.10; cf. *Leviathan* 46, Hobbes 2012, 1052; 1651, 367). *Elements* emphasizes a need for caution since prudence is "nothing else but conjecture from experience, or taking signs of experience warily, that is, that the experiments from which one taketh such signs be all remembered; for else the cases are not alike, that seem so" (EL I.4.10).

Hobbes calls the linkages between conceptions appearing one after another in sense "traynes" of imaginations. Just like all conceptions, either in part or in whole, originate in sense, so also trains—the "Transition from one Imagination to

---

[27] Hobbes provides a similar example in EL I.IV.7 of seeing instances of some offence followed by a particular punishment. Laird (1968, 144) discusses this example and interprets Hobbes's claim that "the foresight of things to come, which is Providence, belongs onely to him by whose will they are to come" (2012, 44; 1651, 10) as identifying God as the only entity capable of such predictive abilities. Chapter 3 discusses how Hobbes's encouragement to imitate God's creative act, and thus acquire *scientia*, is meant to allow humans to possess just this: from causal knowledge one can bring some effect about and can thus predict the results with certainty.

another"—result from the sequential occurrences in sense (2012, 38; 1651, 9; cf. EL I.V.1). These trains reside in memory and can be revisited. Sometimes revisiting a train of imaginations is unguided by passions, like in cases where someone is "not onely without company, but also without care of anything" (2012, 38; 1651, 9). Hobbes identifies these unguided trains as "wild-ranging of the mind," but this does not mean that the linkages themselves are random. Hobbes's example in *Leviathan* of an unguided train shows that his claim that trains of imaginations originate in sense needs clarification. The example reports how someone "in a Discourse of our present civill warre" came to wonder about the value of a Roman penny (2012, 40; 1651, 9). Two trains from experience allow this connection: [from] PRESENT + ENGLISH + CIVIL + WAR [to] KING + DELIVERED and [from] CHRIST + DELIVERED [to] 30 + PENCE [to] COST + ROMAN + PENNY. Clearly, these trains have never occurred together in occurrent *sense* for any human, and Hobbes's reader would only have acquired the latter train through the reading or hearing of Scripture.[28] So Hobbes's claim that successions of one conception to another in trains originated in sense assumes the following: first, that conceptions have decayed; and second, that similarities among them are judged automatically when the mind wanders, like the judgment of similarity that allowed moving from the train containing KING + DELIVERED to the one with CHRIST + DELIVERED.

Hobbes identifies other trains of ideas as "*regulated* by some desire, and designe" (2012, 40; 1651, 9), which he subdivides as follows: (1) trains that seek some cause from an imagined effect and (2) trains that from "imagining any thing whatsoever, wee seek all the possible effects, that can by it be produced" (2012, 40; 1651, 9). Since Hobbes claims that humans share the former trains in common with all animals,[29] he cannot mean 'cause' in the strict sense of knowing *actual* causes that yields *scientia*.[30] Instead, he means that in trains regulated by desire humans and

---

[28] Miller (2011, 28) suggests that this example may be due to Charles I's own comparison between himself and Christ in *Eikon Basilike*, which was published just two years before *Leviathan* and would have been familiar to Hobbes's readers.

[29] Peters incorrectly claims curiosity is responsible for prudence and thus asserts that "Man [. . .] is alone capable of the other kind of regulated thinking which Hobbes called prudence" (1956, 115). Against Peters, Hobbes clearly thinks 'prudence' is not a difference maker: "Neverthelesse it is not Prudence that distinguisheth man from beast. There be beasts, that at a year old observe more, and pursue that which is for their good, more prudently, than a child can do at ten" (2012, 44; 1651, 10–11).

[30] In other words, I am suggesting that this usage of 'cause' is loose and should be understood as seeking antecedents rather than seeking causes proper. This understanding finds support in Hobbes's definition of 'cause' as *causa integra*, according to which a cause is the aggregate of all the accidents of an agent and a patient such that an effect necessarily and simultaneously follows when all are present, since animals (and the *merely* prudent) do not have knowledge of necessary causes but only of antecedents frequently followed by consequents (see *De corpore* IX.3; OL I.107–108; EW I.121–122). Furthermore, Hobbes's claims elsewhere that only a maker knows the causes of an effect and that thus the causes of natural phenomena are underdetermined weighs against thinking of non-human animals as possessing knowledge of actual causes (discussed further in chapter 3, section 3.1). Indeed, in the 1668 Latin version of *Leviathan* 6 Hobbes carefully distinguishes these two types of regulated trains by the different passionate bases responsible for them and removes the ambiguity present with the 1651 English 'knowledge' by using *scire* and *scientia* in contrast to *cognoscere*. He argues that latter kind of train is driven

30   HOBBES'S TWO SCIENCES

other animals examine linkages between conceptions to try to learn, and thus bring about, antecedents with favorable results or avoid those with harmful results. For now, I discuss only regulated trains shared between human and non-human animals and delay discussion of the latter regulated trains—those that are uniquely human—to the account of *scientia* in chapter 3.

When every time in sense the conglomerate conception COYOTE-HOWL + NIGHT has arisen CHICKENS + KILLED has followed, I develop expectations, and I make conjectures when in sense one of the antecedent conceptions occurs. Importantly, Hobbes assumes that the "decay" of motions into imagination and memory, which precedes prudence, will be beneficial to knowers. Were there no such decay, and as a result I remembered every excruciating detail about past nocturnal coyote experiences, to the point of the length of fur, precise shape of ears, and precise eye color, prudence would be impossible to develop and often useless. I would be overwhelmed by the details of my past experiences and fail to make useful conjectures; it matters not, for example, whether the coyotes that have eaten my chickens in the past have had fur 2.35 inches or 2.5 inches in length. To develop prudence I need something general and easier to use, something like the following train of imaginations: [from] COYOTE-HOWL + NIGHT [to] CHICKENS + KILLED.[31] Given such a train having *usefully*-decayed conceptions within it, which represents not length of fur, shape of ears, or eye color, and given my desire to avoid the loss of a food source, a defeasible conjecture can be made: the appearance of the conception CHICKENS + KILLED is imminent. Hobbes calls connections like these "signes" that knowers possess when observing "the Event Antecedent, of the Consequent; and contrarily, the Consequent of the Antecedent, when the like Consequences have been observed, before" (2012, 44; 1651, 11). The more one accumulates "signes," the greater one's prudence is. A prudent chicken farmer immediately runs outside when an antecedent conception appears and will furthermore try to prevent instances of that conception from ever occurring, such as by fencing a larger portion of land or acquiring a large dog to stand guard.

---

by the uniquely-human passion curiosity, which he identifies as a desire "to know (*sciendi*) why and how" and "which is a Lust of the mind, that by a perseverance of delight in the continuall and indefatigable generation of knowledge [1668: *generationis scientiarum*], exceedeth the short vehemence of any carnall Pleasure" (2012, 86–87; 1651, 26; 1668, 28). This unique passion sets humans apart from non-human animals, "in whom the appetite of food, and other pleasure of Sense [...] take away the care of knowing causes [1668: *causarum cognoscendarum*]" (2012, 86–87; 1651, 26; 1668, 28). The second sense of 'knowing causes' just is knowing of antecedents that accumulates to prudence for experienced humans and non-human animals alike. In chapter 5, I argue that to cultivate natural philosophy into 'philosophy' and not mere seeking of antecedents, one must borrow causes from geometry.

[31] I continue using all-caps since so far language has not been introduced. Even using single letters, such 'A' or 'B', to represent these conceptions, which would certainly ease the discussion of them, would involve using a name (as Hobbes notes in *De corpore* II.4 regarding the use of single letters in geometry; see Hobbes 1981, 199; OL I.14; EW I.16).

Likewise, prudent chickens have developed expectations and may attempt to escape from a predator by running or hiding when an antecedent conception arises.

2.3 Classifying Sense-Based Knowledge

Hobbesian prudence clearly counts as a type of knowledge, but since it is grounded in conceptions received in sense it is fallible and Hobbes holds that it must be used "warily." Although he sometimes uses terms inconsistently (see fn. 3 in chapter 1), Hobbes frequently uses the term *cognitio* to describe knowledge based upon sense and accumulated experiences, and he distinguishes it from *scientia*, or knowing scientifically. He distinguishes these two kinds of knowledge in *De corpore* VI.1 as follows:

> We are said to know [*scire*] some effect when we know what its causes are, in what subject they are, in what subject they introduce the effect and how they do it. Therefore, this is the knowledge [*scientia*] τοῦ διότι or of causes. All other knowledge [*cognitio*], which is called τοῦ ὅτι, is either sense experience or imagination remaining in sense experience or memory. (Hobbes 1981, 287–289; OL I.58–59; EW I.66)

So Hobbes divides all knowledge into these two parts: *scientia*, for which we possess knowledge of the causes (i.e., knowledge of the 'why'), and *cognitio*, for which we have sense experience or the remnants of sense experiences in the form of imaginations or memory (i.e., knowledge of the 'that').[32] Human and animal minds can both acquire *cognitio*, but *scientia* is a uniquely human achievement.

Although it lacks the certainty characteristic of *scientia*, prudence will clearly be useful for those who possess it. A prudent individual will reliably conjecture what will happen in the future and, even if they sometimes make mistakes, they will be more likely accurately to predict an outcome than not. Hobbes recognizes that prudent individuals who lack *scientia* are better off than those who make mistakes in reasoning or who "by trusting them that reason wrong, fall upon false and absurd generall rules" (2012, 74; 1651, 21). What sets apart *scientia* from *cognitio* then if the latter, especially when developed to the point of prudence, is beneficial for its holders? Answering this question requires considering an assumption Hobbes makes regarding the role of decay in his philosophical psychology, and the role language plays, before turning to *scientia* in chapter 3.

---

[32] Despite numerous criticisms of Aristotelian philosophy, Hobbes agrees with Aristotle that to have scientific knowledge one must have causal knowledge (for example, OL I.59; EW I.66; OL I.72–73; EW I.82–83; OL IV.42; OL V.156).

Hobbes makes an obvious assumption in his account of prudence: human and animal minds have an innate ability to group similar conceptions together. Pettit calls this the "passive association of ideas" (2008, 15). Hobbes offers no justification for this assumption, but I suggest that he makes a necessary assumption prior to this: the decay of motions from sense to memory automatically decays in a way that is advantageous to human and non-human animal knowers and that is consistent over a series of experiences. Why assume, for example, that particular experiences of coyotes attacking chicken coops will each "decay" in the *same* way so that my association in memory encodes the following: NIGHT + CANINE-SOUNDING-HOWL is frequently followed by CHICKENS + KILLED? What prevents some instances of decay from resulting in NIGHT + SOUND and CHICKEN-COOP-DOOR + OPEN? Or other decay resulting in NIGHT and DOOR + OPEN? While these latter two cases of decay would *sometimes* be advantageous, they would frequently lead to wasted energy since clearly not all sounds in the night are made by predators seeking to dine on poultry, nor are all nights instances in which doors are open. Clearly, prudence could not develop without advantageous and consistent decay of the motions from sense.[33]

What could underwrite Hobbes's assumption that this decay will be both advantageous and consistent and thus allow the prudent individual to predict a consequent conception will likely arise after an antecedent conception has appeared? In short, Hobbes provides no defense of it and his philosophical psychology appears to lack the resources to provide a justification. What seems just as likely as advantageous decay is that some people's decay will be less than advantageous or even harmful such that their memories lack crucial details that relate to their own interests. Indeed, even Hobbes's own analogy of the decay of the motions from sense to waves in a body of water would belie assuming such consistency in decay because there will be interruptions in wave patterns in any body of water due to phenomena like water current patterns, aquatic life, and obstructions like rocks or islands. I suggest that this undefended, and problematic, assumption reveals Hobbes's dependence upon an ability to generalize automatically that humans share with other animals. Without such a bootstrapping move, each conception would remain always radically particular and the sharp details from experience would prevent the grouping of conceptions by salient similitudes (discussed in subsection 2.4.1). Typically, Hobbes assumes that the decay of motions works for knowers' benefit and results in certain features of sense not being present in memory (a predator's length of fur, unique shape of ears, and precise eye color), just like the details of the experience of a city fade as time passes. However, when un-experienced antecedents occur, those with mere prudence may find themselves

---

[33] McIntyre (2021, 99) argues that since decay results in information loss it can result in error. Nevertheless, *pace* McIntyre, Hobbes assumes that for the most part this decay is beneficial even if sometimes useful information may be lost.

making inaccurate predictions or being unable to predict at all. As mentioned already, Hobbes states such scenarios may result in "error," even apart from the usage of names, and "errors of this kind are common to all things endowed with senses" (Hobbes 1981, 271; OL I.50; EW I.56).

2.4 Language and *Cognitio*

An account of the role of language for Hobbes could approach the topic by answering one of the following questions: (1) given the usage of universal names in human language, what is the nature of universal names?; or (2) given the epistemic foundation of particular conceptions in sense, how is the inference from particulars to all of some kind licensed and then aided by universal names?[34] The first aims to explain what universal names do in language (e.g., they denote infinite particulars), but the second elucidates how universal knowledge is possible in the first place (e.g., what inferences are made to arrive at it) and how language aids it. I take the second approach because doing so respects Hobbes's own treatment of epistemic issues first in all of his major works, and furthermore recognizes that the type of knowledge for prudence (*cognitio*) is something shared with non-human animals, which lack language, and so *a fortiori* that epistemic basis must precede language.

Humans and non-human animals can be more or less prudent, and Hobbes admits that some non-human animals are more prudent than humans (2012, 44; 1651, 10–11; also see fn. 29). Once usefully-decayed conceptions accumulate over time, humans notice similarities among them and use language as a tool to augment their natural abilities, which "distinguish[es] men from all other living Creatures" (2012, 46; 1651, 11; cf. EL IV.1). In *Leviathan* 4, Hobbes differentiates what he calls the "general use of Speech" from four "[s]peciall uses of Speech" (2012, 50; 1651, 12–13). The remainder of this chapter will discuss the account Hobbes provides of the general use of speech as it relates to *cognitio*, and two of the "special" uses will be addressed later when discussing the language of discovery that helps in recalling *scientia*. Although both the general and special uses attribute the same basic function to *language*, after discussing how Hobbesian *scientia* is gained through a mental act of considering driven by curiosity in chapter 3, I argue that language plays a unique role in aiding human remembering and demonstrating something *that was newly discovered* from a *single* particular conception (unlike its general use in recalling similarities judged in repeated experiences of particulars).

---

[34] The literature on Hobbes's understanding of language is extensive. Van Apeldoorn (2012) in particular discusses how language aids prudence. For discussion of issues in Hobbes's philosophy of language, see also Hungerland and Vick (1981); Martin (1953); Pettit (2008); Soles (1996).

34 HOBBES'S TWO SCIENCES

Agreeing with Abizadeh (2017, 27) and Watkins (1973, 99), Hobbes held that language arose from private use.[35] In both *Elements* and *Leviathan*, Hobbes attributes the origin of language to the passion of curiosity (EL I.IX.18; 2012, 68; 1651, 20). God's instruction to Adam to name the animals provided the impetus, and the use of names allowed Adam to "register [his] thoughts, [and] recall them when they are past" (2012, 48; 1651, 12; cf. EL I.V.2). Furthermore, names allowed Adam and others to "declare" those thoughts to others. These two uses are subsumed under Hobbes's "generall use of Speech": humans use "*Markes*, or *Notes* of remembrance" privately and "Signes" to show to one another "what they conceive, or think of each matter; and also what they desire, feare, or have any other passion for" (2012, 50; 1651, 12–13). Hobbes views the invention of speech as more monumental than the invention of the printing press, rightly so given the dependence of the latter upon the former, and scholars have placed much emphasis upon his view of signs. This focus is unsurprising since without public signs, he admits, there would be "neither Common-wealth, nor Society, nor Contract, nor Peace, no more than amongst Lyons, Beares, and Wolves" (2012, 48; 1651, 12). The ability to move beyond *natural* prudence, furthermore, hinges on speech, as Hobbes notes: "Those other Faculties [beyond sense and *cognitio*] [...] proceed from the invention of Words, and Speech" (2012, 44–46; 1651, 11). However, Hobbes accords significant importance to private marks, seeing the nature of a name to be "principally" in serving as a mark (OL I.13; EW I.15). Indeed, with marks alone Hobbes declares that "[...] a solitary man can be a philosopher without a teacher; Adam could have been one" (Hobbes 1981, 311–313; OL I.70–71; EW I.80).

Given the importance of marks, I begin by linking Hobbes's discussion of marks to the account of conceptions provided already. As in the *Leviathan* description, in *De corpore* II Hobbes attributes the invention of marks to a need to remember past conceptions and the connections among them: "So fluctuating and frail are the thoughts of men, and so fortuitous is the recovery of them, that the most indubitable experiences can be lost to anyone" (Hobbes 1981, 193; OL I.11; EW I.13). Hobbes likewise accords importance to marks in *Elements* as the means by which we organize our "remembrances" (EL I.VI.1). Hobbes allows for a personal mark to be any sensible object, such as a rock left behind in a certain place to warn of an upcoming danger at sea (EL I.V.1). Instead of spoken words, humans could have used different colored rocks as personal marks and then also as public signs. For example, I could leave myself a pile of rocks in front of a chicken coop to remind myself that a coyote may strike this evening.

Hobbes argues that humans employ conventional signs from their will (*arbitria*) like making a row of stones to remember their property line (Hobbes 1981, 197; OL I.13; EW I.15). If I make such a row of stones for my own use, such as helping

---

[35] Hungerland and Vick (1981) argue against this view. See Macdonald Ross (1987) for cogent criticism of Hungerland and Vick.

PRUDENTIAL KNOWLEDGE FROM SENSE AND THE MECHANICAL MIND   35

myself remember where my property ends, then that sensible monument would serve as a mark; if I do so to keep my neighbor or others off my property, then I have made a public sign. Names in human speech are capable of being both private marks and public signs (*De corpore* II.3), and Hobbes is consistent across multiple works that marks signify an individual's conceptions not things themselves.[36] Hobbes says straightforwardly in *De corpore* II.5: "[. . .] names ordered in speech are signs of conceptions, [and] it is obvious that they are not signs of things themselves" (Hobbes 1981, 201; OL I.15; EW I.17). Hobbes uses the term 'sign' in *De corpore* when discussing both language and the natural world. Similar to the way that natural signs portend *likely* future events, such as "a dense cloud is a sign of consequent rain" (Hobbes 1981, 195; OL I.12; EW I.14), he uses 'sign' to refer to the conventional signs humans use. Since Hobbes holds that humans can never possess causal knowledge of natural phenomena, one could never know that a natural sign has a necessary and causal relation to the consequent event. God as maker knows actual causes and thus knows which signs are necessary and causal (if any are). This issue of knowing through the causes is discussed in chapter 3, and the issue of the suppositional status of human knowledge in natural philosophy is discussed in chapter 5.

Hobbes's view of the private origin of language in "marks" does not fall prey to Wittgenstein's worries because Hobbes does not hold that such a private language could be known only to a lone Adam and thus, in principle, unintelligible to others. In other words, seeing marks as private need not commit Hobbes to what Wittgenstein meant by 'private'. Instead, Hobbes could hold that private use becomes public by sharing the arbitrary marks one uses with others through ostension so that a shared understanding can be reached. Hobbes doesn't say exactly how this would be negotiated between two lone humans at the genesis of shared language through signs (perhaps to my neighbor I point at the pile of rocks in front of my chicken coop in the presence of a coyote approaching at night?), but the goal through such ostension would be "understanding" so that both parties in such an exchange "hath those thoughts which the words of that Speech, and their connexion, were ordained or constituted to signifie" (2012, 62; 1651, 17). Since the conceptions (images) in such an experience will be shared between humans, it may take some back and forth to line up each arbitrary mark correctly so that it can function as a sign, but Hobbes takes such interactions to be possible and, ultimately, responsible for public signs.

---

[36] Sometimes Hobbes's wording appears to belie this straightforward claim that names signify conceptions and not bodies. Duncan (2011) has argued that Hobbes offers a dual view, but see Abizadeh (2015, 10, esp. fn. 48) for criticism. I take Hobbes's account to imply that names signify conceptions and conceptions are caused by bodies, and I leave the details of the further debate to the side. For additional discussion, see de Jong (1990, 67, 73–75), Hungerland and Vick (1981), Laird (1968, 147), and Macdonald Ross (1987).

36   HOBBES'S TWO SCIENCES

The payoff for using a private mark to signify one's conceptions is that it enables the transfer of "Mentall Discourse, into Verbal; or the Trayne of our Thoughts, into a Trayne of Words" (2012, 50; 1651, 13). However, the "transfer" enabled by language does not reflect the structure of particular trains themselves insofar as these trains are composed simply of conceptions in the sequential ordering from particular instances of sense.[37] Were this the case, constructions of marks joined together would be something like transferring the train [from] PRESENT + ENGLISH + CIVIL + WAR [to] KING + DELIVERED to the sentence 'present time' [. . .] 'English Civil war' [. . .] 'King delivered', which would offer no obvious advantage over simply remembering the conceptions themselves.

The issue is not that marks are simply tools for remembering more easily or more reliably. More importantly, the *judgments* of similitudes that Adam made could not be remembered in trains of images (or sequences of marks reflecting their structure), so judgments concerning the trains must also be recorded in marks. Unlike trying to remember by means of images alone, the use of names as private marks enabled lone Adam to remember how he judged those images and for what reason he grouped them with other (judged similar) images. These judgments would frequently be recorded in universal name, such as 'animal' as a mark, which would call to Adam's mind a conception of this or that particular animal; as Hobbes puts it, "any one of those many" (2012, 52; 1651, 13).

2.4.1 Step 1 of Transferring Mental Discursion to Language: Signifying with Marks and Signs

How then does this transfer from trains of imaginations to language work? The distinctions among types of marks that Hobbes offers enables the first step for this transfer. A name need not be a single vocable, as Hobbes notes that the name 'just' is equivalent to the name "*Hee that in his actions observeth the Lawes of his Country*" (2012, 52; 1651, 14), and the name 'Homer' equivalent to 'he who wrote the Iliad' (Hobbes 1981, 205; OL I.17; EW I.19). Names such as the latter are proper, as are 'Peter' and 'John', while others are common like 'man' or 'tree'. The most common of common names are called universals. Hobbes is clear in all of his major works that only names are universals: "[. . .] the name 'universal' is not the name of something existing in nature, nor of an idea or of some phantasm formed in the mind,

---

[37] Against Dawson (2007, 17ff), who connects Hobbes to a tradition of language as a reflection of the mind, in *Anti-White*, as mentioned below, Hobbes mocks Thomas White for holding such a view (see Hobbes 1973, 126; 1976, 53). Normore differentiates Hobbes from sixteenth-century *Nominales* due to this lack of a structural linkage between trains of ideas and spoken language, something shared in common with Ramus's account (2017, 135 fn. 29).

PRUDENTIAL KNOWLEDGE FROM SENSE AND THE MECHANICAL MIND   37

but is always the name of some vocal sound or name" (Hobbes 1981, 205–207; OL I.17–18; EW I.20).[38]

Just as Hobbes does not engage the epistemic skeptic, as argued already, he likewise does not devote much attention to mounting an attack *against* a realist view of universals. The similitudes that a knower judges are simply features of conceptions made salient by their desires, whether to acquire things like food and shelter or to avoid some harm (more below on desires). If humans lacked a fear of death, they may never judge COYOTE as a similitude. A realist about universals might reply that if two or more things are similar in some way, they must be similar with respect to something, such as being a predator or being an animal. Each of those things has the same property, but the instance of each property is different from the others. But, so goes the realist, multiple things having one and the same property presupposes that there are universals, and thus before any talk of judging similitudes can occur it seems that Hobbes needs to give an account of universals as such.

Hobbes's view rejects this presupposition and instead begins from the starting point that contained within conceptions received in sense experience are many different features. Many of those features never matter to the typical human perceiver and will never be noticed. Hobbesian human perceivers are not seeking to gain knowledge about the "objective" features of the world (or universals). Instead, they are embedded within the world of experience themselves and are interested in knowing how to gain what they desire and avoid what they do not desire. As discussed in chapter 3 when introducing what I call the perspectival principle, Hobbes is clear that "[k]nowledge is for the sake of power [*Scientia propter potentiam*]" (Hobbes 1981, 183; OL I.6; EW I.7). Since we "compute only our phantasms," the similitudes that Hobbes describes humans "judging" always begin for perceivers as parts of complex ideas from sense. The particular parts that are judged as similar will depend not only on the complex ideas themselves but also on what individual human beings desire. Given the interplay between complex ideas, which have indefinitely many properties and combinations of properties, and the role that human desires play, Hobbes's view seems to hold that it does not make sense to say there are non-perspectival things that we could intelligibly say exist (i.e., real universals). All that human perceivers can intelligibly posit are complex ideas from sense which are considered in diverse ways depending on the varying desires of perceivers.

Beyond his discussions of names as the only universals mentioned already, this lack of engagement with a realist view is present in his account of the principle

---

[38] For similar claims, see also EL I.V.6, *Leviathan* 4 (2012, 52; 1651, 13), and *De corpore* V.8 (Hobbes 1981, 277–279; OL I.53–54; EW I.60). Although he clearly and consistently holds that there are no universal conceptions in the mind, Hobbes describes "our simplest conceptions" in *De Corpore* VI as universals by which we know the causes of things. I discuss these conceptions in chapter 4.

38 HOBBES'S TWO SCIENCES

of individuation in *De corpore* XI. After criticizing accounts that ground identity claims in matter, form, or accidents, Hobbes recognizes that the principle of individuation should not "always be taken from matter alone, or from form alone" (OL I.122; EW I.137). What similitude one judges in conceptions and then appeals to for an identity claim depends upon one's interests and goals. For example, if I want to know whether Socrates is the same man today as he was fifty years ago, I should not appeal to BODY since "one body always has one and the same magnitude," but I could appeal to FORM, that is, to form that is understood as "the beginning of motion" (OL I.122; EW I.137). Insofar as it is possible to trace the beginning of the type of motion that makes LIVING + HUMAN + BODY in the case of Socrates, recorded as starting at birth or conception and continuing while he is alive, then I can make the identity claim that Socrates today is the same human as fifty years ago even if his matter has undergone many significant changes. But what does it mean to be an *instance* of LIVING + HUMAN + BODY? For Hobbes there need be no universal of which the particular Socrates participates. To identify a similitude in FORM is just to notice bodies moving similarly in a way that correlates with a human knower's desires. In other words, it is only judged similar because the knower cares about knowing whether Socrates is the same now as before, and more generally, only because a knower antecedently cares about the behavior of HUMAN + BODY.

A lone individual like Adam, Hobbes's would-be philosopher, judges similarities among particular conceptions while accumulating experiences (and memories), and on account of those similarities he imposes a universal name: "One Universall name is imposed on many things, for their similitude in some quality, or accident" (2012, 52; 1651, 13). When Adam would use the universal name 'animal' as a mark, it would call to his mind a conception of this or that particular animal, as Hobbes puts it, "any one of those many" (2012, 52; 1651, 13). Conceptions with a judged similarity to one another are marked with what Hobbes calls a positive name that is "imposed because of the similarity, equality or identity" observed among conceptions, and negative names mark "diversity, dissimilarity or inequality" (Hobbes 1981, 203; OL I.16; EW I.18–19).[39] Hobbes connects a further distinction—that between simple and composite names—in *De corpore* II.14 to the earlier example of seeing something at a distance "obscurely" and receiving the conception BODY. With the assumption that the person under consideration has developed sufficient experience and can thus judge similarities among conceptions, as BODY is received in that instance of sense, the simple name 'body' marks that this particular conception shares a similarity, i.e., representing something extended, with other conceptions (Hobbes 1981, 213; OL I.21; EW I.24). Likewise, as ANIMATE and RATIONAL are received, so too each of those

---

[39] Macdonald Ross (1987, 41) convincingly criticizes Hungerland and Vick's attempt (in Hobbes 1981) to interpret Hobbesian "simulitudes" as real universals.

conceptions is marked with a name until the composite name 'animate rational body' is formed, which is equivalent to the name 'man' (Hobbes 1981, 212–214; OL I.21–22; EW I.24). Adam's use of 'man' will bring to mind a conception of this or that man (a conception of a MAN) on account of that particular conception possessing the similitude marked by 'man'.

None of this implies that Hobbes held that with the advent of language *cognition* itself became primarily linguistic or, as Pettit has argued, that according to Hobbes "speech is the source of the capacity to think" (2008, 29ff). Hobbes emphasizes in *Elements* and in *Leviathan* that the orderly succession of conceptions, not names or marks, constitutes "*Mentall Discourse*," stating that he had intentionally used the term 'discursion' in *Elements* to avoid equivocating these two (EL I.IV.1; see also 2012, 38; 1651, 8).[40] Different names for one and the same thing, such as 'just' or 'strong', are due to the fact that there are many "*conceptions* of one and the same thing" (EL I.V.5; emphasis added).[41] When distinguishing between simple and composite names in *De corpore*, Hobbes further emphasizes that composite names follow the composition of conceptions: "[. . .] for as one idea or image is added to another in the mind, and another to this, so to one name is added another and another again, and one composite name comes from all of these" (Hobbes 1981, 215; OL I.22; EW I.24). Indeed, in *Anti-White* Hobbes mocks Thomas White's view that words "reflect the mind":

> "Words," he says on his page 32, "reflect the mind." (We note, by the way, that this is utterly ridiculous, for what resemblance can there be, pray, between a word and the mind? And how is it that, if "words reflect the mind," the languages of all nations are not alike, as their minds are?) (Hobbes 1973, 126; 1976, 53)

Here Hobbes explicitly rejects the view of the mind that Descartes attempts to attribute to him in the Replies to Hobbes's Objections to the *Meditations*. There Descartes wonders "[w]ho doubts that a Frenchman and a German can reason about the same things, despite the fact that the words that they think of are

---

[40] Although Hobbes uses different terms in *Elements*, in *Leviathan* he uses 'discourse' for both mental and linguistic activity.

[41] Hobbes similarly claims that names such as 'pleasure', 'love', and 'appetite' are "divers names for divers considerations of the same thing" (EL I.VII.2). Caution must be exercised when using universal names because they "are not always given to all the particulars, (as they ought to be) for like conceptions and considerations in them all" and, as a result, there is a risk of equivocation (EL I.V.7). Hobbes's example is the name 'faith', which is sometimes used synonymously with 'belief' but other times as only Christian belief, and further still as simply the keeping of a promise. When using a name like 'faith', communicators risk being misunderstood unless they ensure not only that the same *conception* arises for the hearer as the speaker but also that the same *consideration* of that conception arises—this enables understanding (EL I.V.8). One can consider a Christian believer, for example, either as being rightly predicated with 'faith' (such as "That person has faith") insofar as they are believers or insofar as they are *Christian* believers. The attribution of the predicate, and whether one speaker has understanding of another, depends upon the conception raised by the names as well as the consideration of that conception.

40 HOBBES'S TWO SCIENCES

completely different?" (AT VII.179; CSM II.126).[42] However, the prioritization Hobbes places on relationships among conceptions reflects ideal language use. Not all cases of speech will have this connection to conceptions, and as a result will be "completely absurd and insignificant" since "no series of conceptions in the mind corresponds to a series of names" (Hobbes 1981, 223; OL I.26; EW I.29–30).

With sufficient experience and memory amounting to the development of prudence, Adam the lone individual is able to mark "similitudes" he judges among his conceptions as those conceptions decay. According to the earlier example, the prudent chicken farmer acquires something like [from] COYOTE-SOUNDING-HOWL + NIGHT [to] CHICKENS + KILLED. Humans and non-human animals are capable of developing such expectations, which I have earlier argued is possible only after a decay that begins after external objects are absent. Desires determine which consequents become salient (e.g., avoiding the loss of a food source), and humans and non-human animals seek to discover antecedents to avoid undesired consequents. The first step of the transfer of mental discursion to language thus collects myriad conceptions and groups them together with a name like 'coyote-sounding howl'.

2.4.2 Step 2 of Transferring Mental Discursion to Language: Forming Propositions

With language Adam can use the common name 'presence of nocturnal predator' to help him remember similarities he has judged among conceptions over time (step 1): the relevant feature that interests him among various conceptions, given his desire to avoid loss of food, is that of being a nighttime predator, and he has noticed that CANINE-SOUNDING-HOWL + NIGHT is relevantly similar to other conceptions with that accident. He can next form a proposition (step 2): Every 'presence of canine nocturnal howls' is [a case of] 'presence of nocturnal predator', or 'All A are B'.[43] The move to a proposition made of copulated

---

[42] Part of the support Pettit (2008, 27–29) marshals to undergird his view that Hobbes saw language as inaugurating "active, classificatory thought" is a claim that Hobbes developed this account in direct contrast to the Cartesian approach. Against such a claim, I have elsewhere contextualized Hobbes's Objections to the *Meditations* and Hobbes's claims about the role of language in *Elements* and *De corpore* (Adams 2014a). For criticism of Pettit's claims related to the priority of language, see Tabb (2014). See also Paganini (2019, 57) regarding curiosity as preceding language.

[43] While criticizing Hungerland and Vick's account (in Hobbes 1981) of Hobbes's philosophy of language, Macdonald Ross states that "Hobbes insists that notes [i.e., marks] are of individual thoughts, not of complete propositions. So when he says we use notes for recording our thoughts, he cannot possibly mean that we are 'finding out that p' before teaching someone else that p, since 'p' would be a complete proposition" (1987, 44–45). To make this claim, Macdonald Ross appeals to *De corpore* II.3, but, *pace* Macdonald Ross, Adam could not have been a lone philosopher if he did not use propositions (with marks, not signs). Macdonald Ross admits that Hobbes asserts that Adam could have been a philosopher, so why he denies that propositions could be composed of marks is unclear. Indeed, since Hobbes elsewhere claims that "[i]n philosophy, there is only one species of speech [. . .] which most call

PRUDENTIAL KNOWLEDGE FROM SENSE AND THE MECHANICAL MIND   41

names shows a level of categorization non-human animals cannot perform: since the proposition 'All A are B' is an assertion not simply that A has always co-occurred with B but rather that A, the subject, is contained in B, the predicate, or as Hobbes puts it:

> A proposition is speech consisting of two copulated names by which the one who is speaking signifies that he conceives the name which occurs second to be the name of the same thing as the name which occurs first; or (what is the same) the first name is conceived to be contained by the second name. (Hobbes 1981, 225; OL I.27; EW I.30)

The formation of a proposition is thus the second step of the transfer of "Mentall Discourse, into Verbal; or the Trayne of our Thoughts, into a Trayne of Words" (2012, 50; 1651, 13).

In *System of Logic* I.V.2, J.S. Mill complains that "[t]he only propositions of which Hobbes' principle is a sufficient account, are that limited and unimportant class in which both the predicate and the subject are proper names" (1974, 91). Mill's examples against Hobbes are when the subject and predicate are extensionally equivalent (e.g., 'Tully is Cicero'); however, this misses that Hobbes claims only that the second name in a proposition names the same thing as the first (not that the first names the same thing as the second). Thus, Hobbes need not hold that the proposition 'Man is animate' is interchangeable with 'Animate is man'. This is why Hobbes explains that the individual offering a proposition *conceives* the predicate as containing the subject and not the other way around. Later, Mill recognizes that Hobbes's understanding of the proposition is "virtually identical" to one that understands propositions as expressing class relationships, such as an individual within a class or a class within another class, but he argues this view reasons fallaciously *hysteron proteron* because to explain a proposition like 'Snow is white' it must presuppose a knower having a category of 'white' (1974, 94). Mill argues that instead knowers "gradually begin to think of white objects as a class, including snow and those other things" (1974, 94). Hobbes would have no problem granting this latter point, just as was shown that in the case of 'sense' he understands it to include judgment in actual practice, and he would likely respond that conceptual containers such as 'white' that we use as predicates are enlarged as we gain more experience and as our categorizing of accidents is shaped by our desires.

McNeilly argues that Hobbes's account of propositions is "bristling with difficulties" because he claims that either Hobbes assumes that two distinct images are raised in the mind when a proposition is considered or Hobbes assumes that a single image is raised in which different features are considered (1968, 44–45).

'proposition'" (Hobbes 1981, 225; OL I.27; EW I.30), were Adam a philosopher he would have employed propositions with marks.

42   HOBBES'S TWO SCIENCES

McNeilly suggests that if Hobbes agreed to the first option his view faces a worry about identity: how would someone know that these distinct images are of the same thing? If Hobbes agreed to the second option, McNeilly argues that Hobbes must abandon his view that each conception is "determinate," i.e., particular. Against McNeilly, Hobbes would deny that two distinct images are raised and insist that the human mind has an ability to pay special attention ('consider as') to one feature at a time of a single image. McNeilly thinks that such a response raises worries for Hobbes's understanding of imagistic conceptions—claiming "just what sort of imaging can possibly do all the jobs that Hobbes requires?"—but the issue here concerns one of the mind's (innate) abilities to consider its conceptions rather than the conceptions themselves. Hobbes simply assumes that the mind can do this (it is a form of analysis and synthesis), just like a human perceiver can examine a static (external) image, such as a painting, and focus on this or that feature to the exclusion of others, similar to zooming in and out of an image. In other words, Hobbes's assumption concerns the mental activity of "considering as" and not imagistic conceptions themselves. McNeilly instead suggests getting Hobbes out of this worry by claiming Hobbes did not hold that conceptions were all particular and that some conceptions, such as MAN, were "indeterminate" conceptions. Definitions, McNeilly claims, should "produce a conception; and that this *conception* is like a painting, which may be used to represent men generally, but is still a painting of some actual or possible determinate individual" (1968, 46). Hobbes would not agree with this suggestion, since he states in various contexts, as mentioned above, that there are no universals in the mind. My account in chapter 3 argues that we *discover* universality by considering some aspect of a *particular* conception ("considering as" universal) and that a definition helps us remember that act of discovery.

The example related to predators above would be a contingent proposition, since, just like Hobbes claims the proposition 'Every crow is black' does not imply that 'If anything is a crow, then it is black' (see Hobbes 1981, 241; OL I.35; EW I.39; cf. fn. 47 below), there may be cases of not-yet-perceived 'presence of canine nocturnal howls' not contained within 'presence of nocturnal predator', e.g., nocturnal howls by a domesticated dog. As mentioned already, in *De corpore* III.7 Hobbes's states that truth and falsity are applied only to propositions (Hobbes 1981, 233; OL I.31; EW I.35). This account of truth might seem contradictory. For Hobbes writes that a proposition is true just in case a containment relationship holds, i.e., the predicate is a name of every thing that the subject names. However, at *De corpore* III.6, Hobbes says that a negative predicate (name) makes a proposition negative; "Man is not a stone" is Hobbes's explicit example (Hobbes 1981, 233; OL I.31; EW I.35). But the proposition "No man is a stone" does not have a negative predicate and so by Hobbes's understanding it would seem to be positive, since 'no' attaches to the subject rather than the predicate. As a result, it might seem that Hobbes is committed to thinking that the proposition "No man is a stone" is false (since

PRUDENTIAL KNOWLEDGE FROM SENSE AND THE MECHANICAL MIND   43

'stone' is not a name for all the things that 'man' names) but "Man is not a stone" is true (given his definition of negative names in *De corpore* II.7; Hobbes 1981, 203; OL I.16; EW I.18). Hobbes resolves such a worry by providing an equivalency in *De corpore* III.15: "negative propositions, whether the particle of negation be set after the copula as some nations do, or before it, as it is in Latin and Greek, if the terms be the same, are equipollent" (Hobbes 1981, 243; OL I.36; EW I.42). On this account, "*every man is not a tree*, and *no man is a tree*, are equipollent, and that so manifestly, as it needs not be demonstrated" (emphasis original; Hobbes 1981, 243; OL I.36; EW I.42). Given this claim about equipollence, "No man is a stone" is a negative proposition just like "Man is not a stone," and thus Hobbes would hold that both are true.[44]

Adam's automatic judging of similitudes is an ability shared in common with non-human animals, for otherwise non-human animals could not develop prudence. Indeed, Hobbes admits that non-human animals can even possess limited understanding of voluntary signs, such as words, and have an imagination raised upon perceiving them (2012, 36; 1651, 8). However, humans like Adam can also understand "[. . .] conceptions and thoughts, by the sequell and contexture of the names of things into Affirmations, Negations, and other formes of Speech" (2012, 36; 1651, 8). Only the marking of conceptions with names enables this (2012, 62; 1651, 17; see also *De homine* X.1, OL II.88, Hobbes 1994b, 37), allowing humans to create and remember categories like 'presence of nocturnal predator', which can be used as containers (predicates) for names that signify conceptions that are, in many other ways, unlike one another. However, in terms of human desires (e.g., avoiding loss of a food source), these conceptions share a similitude that can be marked with an arbitrarily chosen name (OL I.14; EW I.16).[45] The myriad differences among the conceptions that can be placed into such a cognitive container—perhaps conceptions of snakes, owls, hawks, coyotes, wolves, and so on—is so great that the frailty of memory for human and non-human minds, mentioned already,

---

[44] I thank an anonymous referee of this press for pointing out this potential worry for Hobbes's view of negative names and truth.

[45] Some scholars have taken 'arbitrary' to mean that names are chosen with no consideration whatever to the conceptions caused by bodies. This at least dates back to Leibniz calling Hobbes a "supernominalist" (Leibniz 1976, 128) and was echoed in J.S. Mill's assessment in *System of Logic* I.V.3: "Hobbes' theory of Predication, according to the well-known remark of Leibnitz, and the avowal of Hobbes himself, renders truth and falsity completely arbitrary, with no standard but the will of men [. . .]" (1974, 95–96). This oversimplifies the matter, for Hobbes believes that humans do judge similitudes among their conceptions and use marks to signify those connections. The marking process is arbitrary in two ways: first, the particular name chosen could be any vocable whatsoever; and second, the similitude made salient is from the will (*arbitrium*) of the one using the mark. What matters to a given human will differ from what matters to another human; one human might group 'nocturnal predators' together while another might, for reasons due to differing desires, distinguish 'flighted predators' from 'fur-having predators'. Regardless, the motions of external bodies responsible for the conceptions are not determined by human will, and so once humans agree on how the arbitrarily-selected names are used, there can be intersubjective validity. For discussion, see Adams (2014b) and Jesseph (2010, 124). This is in contrast to Pettit's view, according to which humans and non-human animals are "sensitive to perceptual similarity and difference" in conceptions themselves (2008, 14–15).

44   HOBBES'S TWO SCIENCES

cannot keep them grouped together using imagistic conceptions alone. Names functioning as private marks copulated into propositions make this possible.

Hobbes recognizes the varying categorizations that humans can create according to their particular desires in *Leviathan* 4 when describing what is "*Subject to Names*":

> *Subject to Names* [Latin: *Res nominate est,*] is whatsoever can enter into [Latin: *cogitari,*] or be considered in an account; and be added one to another to make a summe; or subtracted one from another, and leave a remainder. (2012, 58–59; 1651, 15; 1668, 17)

Hobbes identifies this process of summing up the consequences as "the act of reasoning" that the Greeks called the syllogism, and he recognizes that "the same things may enter into account for diverse accidents" (2012, 58; 1651, 16). For example, a conception of a particular HAWK may be marked with common name 'hawk' and then, depending on one's interests in asserting a proposition, may be categorized either as contained within 'flighted predator' or 'nocturnal predator'. Thus, Hobbes distinguishes four general ways humans use names: (1) names of matter or body, such as "*living, sensible, rationall, hot, cold, moved, quiet*"; (2) names of some "accident or quality, which we conceive to be in" some thing; (3) names referring to the human sense by which a thing is perceived, such as when we see something and "we reckon not the thing it selfe; but the *sight*, the *Colour*, the *Idea* of it in the fancy"; and (4) names given to names, such as 'every', or names to given to collections of names, such as 'syllogism' (2012, 58–60; 1651, 160).

The first two ways are relevant to the present discussion and relate to Hobbes's distinction in *De corpore* between concrete and abstract names. A name can be more or less abstract, and prudence will tend to work with more concrete names. When we form a proposition using concrete names such as 'Body A is moved', we may next "make a name for that accident" of 'moved' (2012, 58; 1651, 16), i.e., the accident of a particular body being moved, and seek to know what it is that makes something a body at all or what it is to be movable. In doing so, we are "looking in things for the causes of those names" (Hobbes 1981, 227; OL I.28; EW I.32). As Hobbes puts it: "[...] when we see something first here and then there, we say that it is *moved* or *transferred*, and the cause of its name is that the thing *is being moved* or its *motion*" (Hobbes 1981, 229; OL I.29; EW I.32; emphasis original). The abstract name 'motion' that we use, then, "denotes [*denotare*] the cause of a concrete name, not the thing itself" (Hobbes 1981, 229; OL I.29; EW I.32). Although such names are abstract, Hobbes is careful to say they cannot exist apart from matter: "These are called *names Abstract*; because severed (not from Matter but) from the account of Matter" (2012, 58; 1651, 16; cf. Hobbes 1981, 229; OL I.29; EW I.33).

Commenting on Hobbes's distinction between concrete and abstract names, in *System of Logic* I.V.3 J.S. Mill wonders why Hobbes did not go "one step further, and

PRUDENTIAL KNOWLEDGE FROM SENSE AND THE MECHANICAL MIND   45

[see] that what he calls the cause of the concrete name, is in reality the meaning of it; and that when we predicate of any subject a name which is given because of an attribute (or, as he calls it, an accident), our object is not to affirm the name, but, by means of the name, to affirm the attribute" (1974, 96). Inasmuch as Hobbes states explicitly that abstract names like 'motion' denote the cause of the concrete name (i.e., that our conception of some body is that it is in one place and then another), he is saying that the meaning of the abstract name is the motion of thing, i.e., the accident (as Hobbes uses 'denote'; cf. Macdonald Ross [1987] on this as the meaning of 'denote').

Applying this account to the role of language in prudence, say that I form the following concrete proposition: 'Coyote is nocturnal predator'. My assertion that 'coyote' is contained in 'nocturnal predator' allows me then to *consider* the accident 'nocturnal predator' in an account not directly related to coyotes, that is, in other instances of reckoning by adding or subtracting.[46] Prudential reasoning *could* involve the use of an abstract name, like the name 'predatoriety' (analogous to Hobbes's reference to the humorous Ciceronian examples of '*Appietas*' and '*Lentulitas*'), which would denote the cause of 'predator', i.e., predatory actions by some beast or other. However, given the way prudence seeks only to discover antecedents that precede desired consequences, it will not work with the *most* abstract of names such as 'body' or 'motion'. I return to the issue of abstract names and denotation in chapter 3 where I advance my own view of (1) the distinct roles for language for prudence and *scientia* (introduced in section 3.4); and (2) the role of denotation in allowing a private mark to denote "conceptions of infinite singular things" (discussed in section 3.5) so that individuals who construct geometrical figures do not have to repeat their labors.

2.4.3 Step 3 of Transferring Mental Discursion to Language: Forming Syllogisms

The third step for transferring mental discursion into verbal discourse involves reasoning with propositions in syllogisms. Hobbes's examples of successful syllogisms, featured in his extended treatment in *De corpore* IV, begin with a necessary proposition as the first premise ('Man is an animal'). This counts as a necessary proposition because it is "part of an equivalent name" where the complete,

---

[46] Hobbes holds that a single proposition signifies a single idea, e.g., the proposition 'Man is an animal' signifies just one idea considered first as man and second as animal (see *De corpore* V.9; Hobbes 1981, 279; OL I.54; EW I.61), but given that his account of propositions is one of containment, the expression 'Man is an animal' implies hierarchical relations between these ways of considering that *single* conception. In other words, we utter propositions like this only after having repeated experiences of humans and non-human animals and, depending upon our desires, categorizing them with the expression of containment relationships (this addresses Mill's complaint in *System of Logic* IV.3 [1974, 94] discussed above).

46 HOBBES'S TWO SCIENCES

equivalent name (also a necessary proposition) is "*man is a rational living crea-ture*" (OL I.33–34; EW I.38). For example, Hobbes provides the following as a case of a direct syllogism: Man is an animal; An animal is a body; ergo, Man is a body (Hobbes 1981, 257; OL I.43; EW I.48). Hobbes argues earlier in *De corpore* III.10 that the first premise of this syllogism is necessary since "[. . .] no thing can be conceived or imagined at any time, of which the *subject* is a name while the pred-icate is not" (Hobbes 1981, 237–239; OL I.33–34; EW I.38). In other words, any time a conception signified by 'man' is imagined one must simultaneously have in mind a conception signified by 'animal'.

However, Hobbes's understanding of syllogisms as the joining together of propositions also applies to those that would use contingent propositions, such as Hobbes's own examples of the contingent propositions 'Every man is a liar' and 'Every crow is black' (see Hobbes 1981, 239–241; OL I.34–35; EW I.38). Hobbes defines a contingent proposition as one that "at one time may be true, at another time false" (OL I.34; EW I.38), and given the account of Hobbesian prudence outlined above, the propositions formed from prudence would likewise be contin-gent.[47] Reflecting on this shows how language augments humans' natural abilities in developing prudence (*cognitio*). Consider the following example:

---

[47] Hobbes distinguishes contingent propositions like 'Every crow is black' from necessary propositions like 'Every man is a living-creature' by asserting that the categorical and hypothetical forms of the latter are equivalent while those of the former they are not. In other words, 'Every crow is black' is not equivalent to 'If anything be a crow, the same is black' since the latter is false (see OL I.34–34; EW I.39). Laird complains that Hobbes "darkened his theory" by denying this equivalence, claiming that by 'every crow' Hobbes must mean "every observed crow" and by 'every man' "every man, observed or unobserved, past or future" (1968, 153). I agree with Laird that for prudentially-based, contingent propositions ('Every crow is black') Hobbes does mean "every observed crow"; how-ever, Hobbes makes no claim, *pace* Laird, that 'every man' always relates to observed or unobserved humans since he holds that propositions are necessary "when no thing can be conceived or imagined at any time, of which the subject is a name while the predicate is not" (Hobbes 1981, 237–239; OL I.33–34; EW I.37–38). Regardless of whether one will ever observe human bodies again, one simply cannot imagine a human without also imagining something animate or bodily; the same holds for necessary propositions concerning geometric figures, like SQUARE and TRIANGLE (see *De corpore* III.13; OL I.36; EW I.40). These latter examples are the subject of chapter 3. Recognizing that Hobbes's examples of copulation use necessary propositions, Nuchelmans suggests that the "essential func-tion of the copula" is to mark *necessary* connections (1983, 134–135), but Hobbes never denies that contingent propositions use copulas, nor need he. Instead, Hobbes's account implies that contingent propositions express the same relationship between names as necessary propositions do (i.e., that "the first name is conceived to be contained by the second name" [Hobbes 1981, 225; OL I.27; EW I.30–31]), but they do so in a way that the *reflective* user recognizes to be defeasible. Given Hobbes's view that the actual causes of natural phenomena are unavailable to humans, the propositions of natural philosophy are a prominent example of contingent propositions. Wilson (1996) recognizes a role for contingent propositions in Hobbes's explanations of natural phenomena but, in claiming that Hobbes offers natures of bodies like GOLD as "nominal definitions," neglects the role played by necessary ge-ometrical principles (*scientia*) in serving as borrowed (possible) causes. I argue in chapter 5 that this gives the propositions of natural philosophy *suppositional* certainty: we cannot know with certainty that the geometrical principle borrowed is in fact responsible (since we are not the makers of natural things); nevertheless, we can know that *if* it were occurring then some result would necessarily occur and also that *if* that result were involved then the phenomenon in question would necessarily occur. de Jong (1990, 83) connects contingent propositions to Hobbesian "knowledge of fact" and recognizes that such propositions could not be asserted on the basis of a single conception (unlike those of *scientia* discussed in chapter 3). However, de Jong then claims that "Hobbes seems to have missed this point

PRUDENTIAL KNOWLEDGE FROM SENSE AND THE MECHANICAL MIND   47

Let $A =_{df}$ 'presence of canine nocturnal howls'
       $B =_{df}$ 'presence of nocturnal predator'
       $C =_{df}$ 'death of chickens'

       P1: Every A is B
       P2: Every B is C
       Ergo: Every A is C

Unlike Hobbes's explicit examples of syllogisms, each premise in the syllogism above contains a contingent proposition. This syllogism succeeds in being a summing up of three names (A + B + C) into a single sum (Every A is C), as Hobbes indicates in *De corpore* IV.6 is characteristic of syllogistic reasoning (see Hobbes 1981, 255; OL I.42; EW I.48). Furthermore, the syllogism succeeds in Hobbes's view since although it is from contingent premises it is not from two particular propositions, such as a syllogism with premises 'Some man is blind' and 'Some man is educated' (see Hobbes 1981, 255; OL I.41–42; EW I.47).

Such a syllogism, representing *prudential* reasoning as opposed to *philosophical* reasoning that could yield *scientia*, is analogous to one Hobbes provides that uses a singular name, as follows: Some man is Socrates, Socrates is a philosopher; therefore, some man is a philosopher (Hobbes 1981, 253; OL I.41; EW I.46–47). Regarding that syllogism, though it is a valid syllogism, i.e., one in which "when the premises are conceded, the conclusion cannot be denied," it is one that is nonetheless "useless to Philosophy [*inutilitis ad Philosophiam*]" (Hobbes 1981, 253; OL I.41; EW I.46–47). Likewise, the syllogisms formed to represent prudential reasoning, as a type of *cognitio*, will be useless to philosophy, in which one seeks knowledge of actual causes, but they nevertheless will allow humans using names, propositions, and syllogisms to improve upon their natural abilities by using conclusions such as 'Every A is C' to categorize according to their experiences. The prudential syllogism completes the usage of marks that Adam or some other lone individual could use in isolation for prudence.

In addition to serving as private marks, names play a second role: to declare by publicly used words (signs) to others the contents of one's mind. Words as public

---

[that contingent propositions do not follow from a single conception] completely" (1990, 87). Given Hobbes's many references to the epistemic defeasibility of prudence, that the propositions that humans use to register it have a different foundation would be no surprise to Hobbes at all. de Jong holds this because he takes generalizing to be something possible only with language, i.e., a "language-dependent intensional analysis," and that, according to Hobbes, in language one can move from a single *proposition* (not conception) to a universal proposition just like one uses the predicate logic rule of universal generalization from an arbitrarily selected constant. My account in chapter 3 disagrees by showing a lesser role for language; it helps us "register" what we discover by "contemplating" our *conceptions* (not, as de Jong would suggest, by using the rules of predicate logic to generalize from some proposition Ta → Ra, where a is some arbitrarily selected constant, Tx is x is a triangle, and Rx is x has angles that sum to two right angles, to a universal proposition $(x)(Tx → Rx)$).

signs that humans use are thus "to signifie (by their connexion and order,) one to another, what they conceive, or think of each matter; and also what they desire, feare, or have any other passion for" (2012, 50; 1651, 13). Hobbes is explicit that names as marks is the foundation of names as public signs, for what good would names as public signs (without being marks) serve someone who "existed alone in the world" (Hobbes 1981, 197; OL I.13; EW I.15). This additional role played by names as signs informs Hobbes's second version of "understanding" in *Leviathan*. As mentioned earlier, Hobbes recognizes in *Leviathan* 2 that there is a sense in which non-human animals understand, such as the dog who understands its owner's call (see 2012, 36; 1651, 8). However, there is a clear sense in which non-human animals do not possess understanding insofar as they cannot cognize what a human expresses with signs of the "connexion and order [. . .] one to another, what they conceive, or think of each matter" (2012, 50; 1651, 13). Imagine that I utter the proposition 'A treat is a nutritious snack' and next ask the question 'Would you like a treat?' In both utterances, a canine companion has a conception of a particular TREAT come to mind, but it will not detect that in the former utterance I have asserted a proposition in which I categorized 'treat' as contained in 'nutritious snack'.[48] As a result, after discussing the functions of marks and names, Hobbes defines the understanding that is unique to language speakers as "nothing else, but conception caused by Speech" (2012, 62; 1651, 17).

2.5 Conclusion

This chapter has been devoted to *cognitio*, the term Hobbes used to classify knowledge that arises for humans and non-human animals alike from the myriad particulars of sense. The source of *cognitio* is particular instances of sense that cause imagistic conceptions in the mind. Those conceptions linger in the mind, and after repeated sense develop into experience and eventually memory. These imaginations form what Hobbes identified as trains, and to explain human and non-human behavior Hobbes posits *regulated* trains of imaginations that seek to discover some antecedent that preceded a desired consequent. Language helps humans augment their natural abilities so that when they notice similarities among their conceptions they use language, which "distinguish[es] men from all other living Creatures" (2012, 46; 1651, 11; cf. EL IV.1). In the case of private marks, humans can remember salient "similitudes" among their conceptions and categorize them by means of propositions. However, the propositions developed from prudence are contingent, like 'Every crow is black' or 'Every hawk is a nocturnal predator'. Although all prior experiences of a CROW included BLACK, this need

---

[48] For discussion of a similar example, see McIntyre (2021).

not imply necessity that future instances of a CROW will so too. Nevertheless, the reasoning that humans employ by means of syllogizing with contingent propositions allows them to rise above non-human animals and, with public signs, declare the categorizations they have made (by propositions) and communicate the consequences of their reasoning (by syllogisms) to other humans with language.

3

Scientific Knowledge from Making and the Mechanical Mind

According to Hobbes, knowing with certainty is, in the parlance of contemporary epistemology, a special form of knowing-how.[1] This form of knowledge (*scientia*) is possible only for an individual who creates something—the creator knows *how* their creation was made and thus has access to knowledge of its causes. Humans' ability to gain *scientia* results from a capacity to make new conceptions from an abstracted conceptual foundation through a careful process of construction, and Hobbes singles out two disciplines in which this certain knowledge is possible: geometry and civil philosophy. As with the other acts of the mind discussed in chapter 2, Hobbes aimed to make the cognitive act that provides *scientia* intelligible using only bodies in motion (imagistic conceptions taken apart and put together). However, unlike the abilities that underlie *cognitio*, this chapter shows how Hobbesian *scientia* requires great effort (industry), is driven by a uniquely human passion (curiosity), and is possible only in ideal circumstances (when in a state of leisure and when possessing the proper method). Seeing *scientia* as possible only through the causes is something that Hobbes, despite his many criticisms of Aristotelianisms, praised in Aristotle. For example, in *Principia et Problemata* he endorses a "truth" of Aristotle that "knowing [*scire*] is knowing [*scire*] through the cause" (OL V.156), and he places that same claim in speaker B's mouth in *Examinatio* when criticizing John Wallis's view of demonstration (discussed in chapter 5) (OL IV.42).

This chapter articulates two criteria of Hobbesian *scientia*: 1) an instance of *scientia* is knowledge of actual causes; and 2) since *scientia* is power (*potentia*), what counts as *scientia* follows from a knower's aims. Corresponding to these criteria, I argue Hobbes's epistemology of *scientia* is grounded upon two epistemic principles:

---

[1] The contemporary literature on knowing-how, and whether it differs from knowing-that, is voluminous. Gilbert Ryle's (1949) argument that they are distinct types of knowledge has been influential, but there have been recent defenses against the distinction by intellectualists (e.g., Stanley and Williamson 2001). For criticisms and discussion of Stanley and Williamson, see Adams (2009), Koethe (2002), Moffett and Bengson (2007), Noë (2005), and Snowdon (2003). For a recent overview of the issues, see Pavese (2021).

Hobbes's Two Sciences. Marcus P. Adams, Oxford University Press. © Marcus P. Adams 2025.
DOI: 10.1093/9780198924715.003.0003

The Maker's Knowledge Principle: only the maker of something knows its causes

The Perspectival Principle: *explananda* are determined by human interests and aims, and an *explanadum* determines which *explanans* is relevant

In addition to making *scientia* mechanically intelligible,[2] Hobbes's emphasis on interests and aims makes *scientia* a type of perspectivalism. There is no theoretically-privileged "level," as it were, from which one should draw causes. A knower's interests and aims determine what is to be explained—whether how to make peace or a geometric figure—and that end determines what serves as a foundation for construction. Although Hobbes is a materialist, he is no reductionist.

This chapter addresses the source of Hobbesian *scientia* as a particular construction (an inference from the construction of *one* particular to *all* of a kind) and shows Hobbes's explanation for this behavior by appeal to the passion of curiosity (sections 3.1–3.4). Next it examines the role language of language in aiding knowers to mark conceptual discovering (*invenire*) made in construction with private marks, form necessary universal propositions that serve as definitions, and then use those to form syllogisms (section 3.5), and concludes by articulating the way a knower's perspective determines how Hobbes's method of analysis is used and also how causal principles are used in natural philosophy (section 3.6).

3.1 The Possibility of Causal Knowledge

The value of *scientia*, like prudence, lies in its utility. In appealing to the usefulness of scientific knowledge, Hobbes dissolves any distinction between theoretical and practical knowledge. He declares in *De corpore* I.6 that *scientia* provides benefits to its holders: "[k]nowledge is for the sake of power [*Scientia propter potentiam*]" (Hobbes 1981, 183; OL I.6; EW I.7).[3] The goods received from *scientia* include the "comforts" stemming from knowledge of geometry and its uses in physics, such as marking "moments in time" and "mapping the face of the earth" (Hobbes 1981, 183–185; OL I.6-7; EW I.7). Civil and moral philosophy provide benefits because of what they prevent, as Hobbes argues that "civil wars and then the greatest

---

[2] For discussion of making and causal knowledge in Hobbes's geometry, see Jesseph (1996, 88ff). Some connection has been made between Hobbes and others who saw making as essential to scientific knowledge, for example, Bacon and Vico (Barnouw 1980b; Gaukroger 1986). Pérez-Ramos (1989) argues that for Francis Bacon making is the ideal of scientific knowledge, but there are significant differences between this understanding of Bacon and Hobbes. In particular, Hobbes claims we possess maker's knowledge only in geometry and civil philosophy; as a result, he could never countenance, as Bacon does according to Pérez-Ramos, that humans possess maker's knowledge in natural philosophy.

[3] Hobbes describes his geometry of *De corpore* Part III as including only what "[. . .] is new, and conducing to natural philosophy" (OL I.176; EW I.204).

52 HOBBES'S TWO SCIENCES

calamities follow upon ignorance of political responsibilities, that is, moral science [*moralis scientiae*]" (Hobbes 1981, 187; OL I.8; EW I.10).[4]

A qualitative difference between prudence (as a type of *cognitio*) and *scientia* is that the former provides only associations that amount to "*Praesumption* of the *Future*" (2012, 44; 1651, 11), not causal knowledge. Acausal prudence cannot prevent civil wars since such conflicts result from "being ignorant of the cause of war and peace [*bellorum et pacis causae*]" (Hobbes 1981, 185; OL I.7; EW I.8). Hobbes's requirement that *scientia* include knowing the causes is displayed in his definition of 'philosophy' in *De corpore* I.2, which has two paths: one path from actual causes to effects, and another path from effects to possible causes (Hobbes 1981, 175; OL I.2; EW I.3).[5] Taking either path counts as philosophizing, but only the former is the path of *scientia* since it involves moving from actual causes, known only to makers, to effects following from those causes. The latter is the path of natural philosophy (discussed in chapter 5). Unlike these two paths and their appeal to causes, whether actual or possible, acausal prudence develops from accumulated experiences that are "but a Memory of successions of events in times past" (2012, 1052; 1651, 367). While useful in many circumstances, prudence fails in previously unexperienced situations, as Hobbes notes: "[. . .] the omission of every little circumstance altering the effect, frustrateth the expectation of the most Prudent" (2012, 1052; 1651, 367). Even when a prudent individual is in a circumstance nearly identical to past experiences and uses language to remember that appearances of 'A' have always been followed by 'B', prudence cannot license an inference of 'B' with certainty. This lack of certainty explains why Hobbes holds that true contingent categorical propositions, like 'Every crow is black' or 'Every man is a liar', fail to entail hypothetical propositions (Hobbes 1981, 241; OL I.35; EW I.38–39).[6]

---

[4] Hobbes's praise of *scientia* as *potentia* in *De corpore* I.6 may seem contrary to his statements in *Leviathan* 10 that appear to diminish its value: "The Sciences, are small Power; because not eminent; and therefore, not acknowledged in any man; nor are at all, but in a few; and in them, but of a few things. For Science is of that nature, as none can understand it to be, but such as in good measure have attayned it" (Hobbes 2012, 134; 1651, 42). Hobbes alters this in the 1668 Latin as follows: "Science, is power [*Scientia, Potentia est*]; but a small one, because conspicuous science is rare, and therefore is not apparent except in very few people [. . .]" (Hobbes 2012, 135 trans. in fn. 10; 1668, 44). However, Hobbes's exaltation of *scientia* in *De corpore* is consistent with these claims in *Leviathan* because the latter is merely articulating the social nature of *perceived* power. For example, he identifies "Reputation of power" as power (2012, 132; 1651, 41) and recognizes that public perception of power may be misguided, as in cases related to "arts of publique use" (2012, 134; 1651, 42). Skinner (2002, 83) takes Hobbes's identification of science as "small Power" and eloquence as "Power" to indicate that in *Leviathan* Hobbes had come to see the need for civil science itself to include eloquence so that it could persuade; as I argued in chapter 1, however, Hobbes's reason for why eloquence is power—because it is "seeming Prudence" (2012, 134; 1651, 42)—implies that it cannot be part of *scientia*.

[5] Hobbes provides slightly different definitions at *De corpore* VI.1 (OL I.58; EW I.65–66) and *De corpore* XXV.1 (OL I.313; EW I.387). An early definition appears in Robert Payne's notes of *De corpore* (Chatsworth ms. A.10; cf. Hobbes 1973, 463). See also the definition of philosophy and exclusion of prudence from it in *Leviathan* 46 (2012, 1052; 1651, 367).

[6] de Jong (1990, 82–83) neglects the different epistemic bases of *cognitio* and *scientia* and criticizes Hobbes's account of contingent propositions, such as 'Every crow is black', by claiming they could not be asserted as true on the basis of mere particular conceptions. But Hobbes's nominalism requires that

SCIENTIFIC KNOWLEDGE FROM MAKING AND THE MECHANICAL MIND   53

Since prudence results from an accumulation of experiences that form trains of imaginations, Hume's later worries regarding 'cause' in Sect. IV of *Enquiry concerning Human Understanding* would threaten Hobbesian *scientia* if Hobbes thought prudent individuals were licensed to claim 'A causes B' from trains of imaginations like '[from] A [to] B'. However, Hobbes separates *scientia* from *cognitio* and its sensory basis, as mentioned already in chapter 2: "We are said to know [*scire*] some effect when we know what its causes are, in what subject they are, in what subject they introduce the effect and how they do it. Therefore, this is the knowledge [*scientia*] τοῦ διότι or of causes. All other knowledge [*cognitio*], which is called τοῦ ὅτι, is either sense experience or imagination remaining in sense experience or memory" (Hobbes, 1981, 287–289; OL I.58–59; EW I.66). What sets apart this ability to know causes from prudence and the forms of knowing (*cognitio*) identified with sense?

Rather than in accumulated experiences, Hobbes grounds causal knowledge in the act of making with what I have called the maker's knowledge principle. Such making, Hobbes claims, is possible only in geometry and civil philosophy, as he notes in the Epistle Dedicatory of *Six Lessons to the Professors of the Mathematiques* (hereafter *Six Lessons*):

Geometry therefore is demonstrable for the lines and figures from which we reason are drawn and described by ourselves and civil philosophy is demonstrable because we make the commonwealth ourselves. (1656a, sig. A2r-v; EW VII.184)

Hobbes similarly asserts in *De homine* 10.5 that geometry is demonstrable because "the causes of the properties that individual figures have belong to them because we ourselves draw the lines" (OL II.93–94; Hobbes 1994b, 41–42).[7] Regarding "politics and ethics," he likewise claims that they "[. . .] can be demonstrated a priori; because we ourselves make the principles—that is, the causes of justice

---

particulars be the basis for either necessary or contingent propositions. According to the account I offer in the present chapter, the unique foundation of *scientia* (a foundation reached either through analysis of a conception in discovery or begun with a thought experiment in "teaching") removes potentially confounding causes and allows the maker to know with certainty the causes of what they construct from a particular. Both necessary and contingent universal propositions assert containment relationships but, given the different epistemic basis of prudence (*cognitio*), when 'every' is used in a universal contingent proposition it implies only 'every x [that I have observed] has feature y'. See chapter 2, fn. 47.

[7] Pettit puzzlingly describes Hobbes's reference to geometry in *De homine* 10.5 as "the most basic account of bodies, especially natural bodies" and claims that this is an area where Hobbes saw himself as "working in a demonstrative or a priori mode" (2008, 19). While Hobbes does hold that geometry should ideally be *used* in natural philosophy within some explanations, as I discuss in chapter 5, these two projects are importantly distinct (one is *scientia* and the other is ideally, as Hobbes says, a type of mixed mathematics).

(namely laws and covenants)—whereby it is known what justice and equity, and their opposites injustice and inequity, are" (OL II.93–94; Hobbes 1994b, 41–42). Such knowledge is possible—and demonstrable—because "the construction of the subject whereof is in the power of the artist himself, who, in his demonstration, does no more than deduce the consequences of his own operation" (EW VII.183– 184). As will discussed, ultimately the definitions in these two sciences must "register" causes that show how to make the definiendum. Thus, in geometry causal definitions show how to make figures, e.g., 'a line is made by the motion of a point', and in civil philosophy the Laws of Nature show how to make peace.

Some scholars have seen Hobbes's understanding of science as conflicted. For example, Malcom finds two competing views of scientific knowledge in Hobbes's thought beginning in the 1640s period when he was at work on *Anti-White* and later in *Leviathan* and *De corpore*, calling these "the knowledge of causes and the knowledge of definitional meaning" (2002, 17, 154–155). Indeed, sometimes Hobbes defines 'science' by appearing to link it only to knowing the meanings of terms (e.g., the claim that it lies "first in apt imposing of names" in *Leviathan* 5; 2012, 72; 1651, 21; though, cf. *Elements* I.VI.4 where conceptions from sense are the first of four "steps"). However, Hobbes's requirement that definitions in *scientia* be generative removes the need to posit the two "competing views" that Malcolm suggests. This requirement is clear in Hobbes's criticisms of Wallis and Euclid (discussed in chapter 1). In *Examinatio*, speaker A links these two by first bringing B to concede that definitions are "the principles of the sciences" and that all knowledge is from an understanding of the causes. After B assents to these two claims, A leads B to agree that knowledge of the cause must be contained in a definition and, as a result, to hold that those who explain the "generation [*generationem*] of a thing" define it best (OL IV.86–87).

In other words, since definitions in Hobbesian *scientia* must be generative, Malcolm's "competing views" collapse into the same thing. Nevertheless, Malcolm suggests Hobbes's "floundering" was due to attempts to unify his philosophy, and Malcolm claims that this is evident in the definition of 'science' in *Leviathan* 5, which, he states, "compromises awkwardly between the knowledge of the consequences of names and knowledge of the consequences of facts" (2002, 155). Two additional points can be made to reject seeing this as an awkward compromise. First, Hobbes's inclusion of 'fact' in the definition of 'science' in *Leviathan* 5 could be because there he means 'science' in the broader sense of including geometry and civil philosophy (*scientiae* proper) as well as natural philosophy, the latter of which lies between prudence and *scientia* insofar as it can borrow causal principles from geometry and mix those with knowledge of the "subject" (discussed in chapter 5). Second, the textual context of *Leviathan* 5 immediately following the brief excerpt upon which Malcolm focuses makes clear that 'science' is about *doing something* beyond merely observing how 'facts' depend upon one another. Hobbes claims this dependence is that

SCIENTIFIC KNOWLEDGE FROM MAKING AND THE MECHANICAL MIND   55

[...] by which, out of what we can presently do [1668: *facere*], we know how to do something else [1668: *docetur aliquid aliud simile facere*], when we will, or the like, another time: Because when we see how anything comes about, upon what causes, and by what manner; when the like causes come into our power, wee see how to make it produce the like effects [1668: *docemur (quoties in nostra potestate sunt causae similes) similes producer effectus*]". (2012, 72–73; 1651, 21; 1668, 23)

The double use of *docere* in the 1668 Latin edition, rather than perhaps *videre*, or even *cognoscere* or *scire*, emphasizes that when we do something, i.e., *make* something, we are thereby *taught* how to make something similar, how to make similar effects (as discussed below, Hobbes elsewhere frequently uses *docere* to describe demonstration). Before discussing causal definitions further, and the roles played by language in *scientia*, it is necessary to treat the role conceptions play in gaining *scientia* when humans consider conceptions in various ways and construct from those considerations.[8]

Since only makers can have causal knowledge, it follows that only God knows the causes of natural phenomena. Hobbes asserts in *Six Lessons* that the actual causes of natural phenomena are unavailable to humans because they "know not the construction, but seek it from the Effects"; as a result, humans posit only causes that "may be" (1656a, sig. A2v; EW VII.184). Likewise, in *Tractatus Opticus II* (c. 1640) Hobbes notes that humans offer suppositions that may be false:

[...] and since it is not impossible [*non sit impossibile*] that similar phenomena be produced by dissimilar motions; it can be held that from the supposed motion the effect may be rightly demonstrated, even if nevertheless the supposition [*suppositio*] itself not be true. (Hobbes 1963, 147)[9]

Similar phenomena can be produced by different motions, so humans cannot know which motions God used and the actual causes of natural phenomena are underdetermined. Hobbes repeats these claims about humans' inability to know actual natural causes in various other works.[10] The issue of knowing natural causes will be addressed later in chapter 5; for now, I raise it to show how the maker's

---

[8] My account thus shows *how* knowers use causal definitions to "register" their own constructions with conceptions in discovery or to teach others, *pace* Pettit's claim that Hobbesian *scientia* is simply "grounded in the definitions of the terms used, as in Euclidean geometry" (Pettit 2008, 19). Indeed, Hobbes criticizes Euclid for lacking causal definitions of some figures (e.g., EW VII.184).

[9] Alessio (Hobbes 1963) is a transcription of Harley ms. 6796, ff. 193–266. Malcolm (1996, 25) argues that this manuscript, commonly referred to as *Tractatus Opticus* II, "must have been completed by 1640," though it is commonly dated 1644.

[10] For example, see *De homine* 10.5 (OL II.93), *Decameron Physiologicum* (EW VII.88), *Principia et Problemata* (Hobbes 1674, 38ff; OL V.209), and *Seven Philosophical Problems* (1682, sig. Av). Likewise, he distinguishes the project of natural philosophy in *De corpore* Part IV as offering suppositions to account for how effects *may* be generated, not how they are (OL I.316; EW I.388)

56   HOBBES'S TWO SCIENCES

knowledge principle informs Hobbes's understanding of the limits of natural philosophy.

Hobbes encourages the reader of *De corpore* to "imitate the creation," analogizing philosophizing to the Biblical act of creation:

> If you are going to pay serious attention to philosophy, let your reason hover over the confused abyss of your thoughts and experiences. The confused things must be shaken violently, distinguished, and ordered, having been marked with their own names, that is, in method [*Methodo*] it must be according to the creation of things themselves. (Hobbes 1655, *De corpore, Ad Lectorem*, sig. A4v.; OL I. unnumbered; EW I.xiii)

Hobbes's allusion to the Biblical story reflects not a creation *ex nihilo* but a world shaped from what was "formless." Just as the spirit of the Creator God shaped what was already there, Hobbes exhorts would-be knowers to examine the pre-existing contents of their minds, conceptions received in sense remaining as imaginations and memories, and subject those existing contents to intense examination—they must be "shaken violently, distinguished, and ordered." Hobbes uses similar imagery in *Leviathan* when he identifies Nature as "the Art whereby God hath made and governes the World" and claims that humans create "that great LEVIATHAN called a COMMON-WEALTH" by "imitating that Rationall and most excellent work of Nature, *Man*" (2012, 16; 1651, 1).[11] The following sections show how Hobbes thinks this imitation should be done.

3.2 The Mental Act of Considering As for Discovery

A significant problem for humans and other animal knowers addressed in chapter 2 is that sense presents a complex reality. Conceptions received in sense encode more details than will be useful, so Hobbes assumes that a decay of conceptions precedes the formation of the passive associations that enable prudence to accrue with "signes" (2012, 44; 1651, 11). This decay aids the development of prudence by removing features of conceptions received in sense in a way that is advantageous and consistent. Humans then use marks and signs to remember judged similitudes, and they employ contingent propositions to reflect relationships among names and the conceptions they signify.

---

[11] Miller rightly emphasizes that Hobbes's reference to "imitation" in the introduction of *Leviathan* refers to imitating God's *activity* as creator and not merely imitation of what God's art produces (2011, 54). Imitation of God's act of creating provides humans with novel outputs—geometric figures and commonwealths—that otherwise would not exist. On the context of Hobbes's appeal to the imitation of God, see Miller (2011, 59–70)

SCIENTIFIC KNOWLEDGE FROM MAKING AND THE MECHANICAL MIND   57

Instead of judging similitudes among *many* conceptions and remembering them with names and propositions, to acquire *scientia* human knowers must go to a conceptual foundation and from that bedrock construct something. This qualitative difference between the acquisition of *cognitio* and *scientia* is reflected in various works.[12] For example, in *Leviathan* 20 Hobbes describes the "greatest objection" to his account of sovereign power from those who have never seen such a commonwealth: "[. . .] an argument from the Practise of men, that have not *sifted to the bottom*, and with exact reason weighed the causes, and nature of Commonwealths, and suffer daily those miseries, that proceed from the ignorance thereof, is invalid" (2012, 320; 1651, 107; emphasis added).[13] Prudence grounds this objection from "practice" and takes the following form: no other such commonwealths have been observed, therefore one is unlikely to appear or even be possible. Regardless of the usefulness of prudence in everyday life, in the context of the *science* of civil philosophy Hobbes rejects it out of hand and retorts that even if "in all places of the world, men should lay the foundation of their houses on the sand, it could not be thence inferred, that so it ought to be" (2012, 320–322; 1651, 107). Unlike prudence-based practice, which is akin to "Tennis-play," Hobbes argues that the "skill [1668 Latin: *scientia*] of making, and maintaining Common-wealths" consists in "certain Rules," just like those followed in arithmetic and geometry. To discover these rules, Hobbes holds that there must be a combination of leisure, curiosity, and method. Leisure has not been had by most "poor men," and those with leisure and a will (curiosity) to sift to the bottom have not yet had a method to do it.[14] The proper method for the science of civil philosophy, Hobbes holds, is something he has supplied.[15]

This need to "sift to the bottom" to achieve *scientia* is part of the imagery Hobbes uses immediately preceding the extended quotation above in section 3.1 from the Epistle to the Reader of *De corpore*:

---

[12] Additionally, I have already noted another qualitative difference between these two: *scientia* is causal and *cognitio* is not. This need not imply that these forms of knowing have no overlap in their usage (*contra* Deigh 1996, 39–41). Indeed, since Hobbes identifies *scientia* as *potentia* the results of *scientia* will often be employed for prudential aims (e.g., I may desire to avoid war given past associations with lack of peace and thus seek out a political scientist's advice). van Apeldoorn (2012, 160–161) cogently criticizes Deigh's account (1996), arguing that *scientia* is driven by the passion of curiosity and serves human ends. My account shows that Hobbesian *scientia* requires a different cognitive activity from *cognitio*: reaching a foundation from which a construction can be made.

[13] In the 1668 Latin, Hobbes rephrases this to "[. . .] never deeply looked into the causes and nature of civil matters" (2012, 320 fn. 67; 1668, 103).

[14] This understanding relies upon the alteration Hobbes makes in the 1668 Latin (cf. 2012, 322 fn. 70). The English 1651 rendition allows the possibility that there have previously been individuals without *both* the passion of curiosity and a method: "[. . .] which Rules, neither poor men have the leisure, nor men that have had the leisure, have hitherto had the curiosity, or the method to find out" (2012, 322; 1651, 107).

[15] In the Epistle Dedicatory of *De corpore* (EW I.ix), Hobbes describes civil philosophy as having begun with his *De cive* and compares his own ingenuity with that of Galileo and Harvey.

58   HOBBES'S TWO SCIENCES

"[. . .] Philosophy, the daughter of your mind and the whole world, is in you your-self; perhaps not yet fashioned, but similar to the world, her creator, formless, as it was in the beginning. Therefore *you must do what those who make statues do, who, carving out the unnecessary parts, do not make the likeness but discover it*". (Hobbes 1655, OL I unnumbered; emphasis added)

For human knowers to "carv[e] out the unnecessary parts," I suggest Hobbes appeals to a mental act that I introduced in chapter 2 and will develop further here: human knowers have the ability to consider their imagistic conceptions in various ways. To distinguish a special use of this mental act from the broad ability that all human knowers have to consider this or that feature of their conceptions in reasoning, I use 'considering as' to refer to cases when considering is driven by the uniquely-human passion of curiosity for those who possess a method and have leisure.[16]

Most generally, knowers may consider a unified, conglomerate conception of a body in various ways by focusing on certain features and ignoring other features. As argued in chapter 2, for prudence this involves judging similitudes among one's conceptions, which are made salient by one's desires, using a proposition to cat-egorize the names signifying those conceptions (e.g., 'coyote is nocturnal pred-ator'), and then considering those salient features in reasoning (in an "account").[17]

---

[16] As will be discussed in examples below and in subsequent chapters, Hobbes describes mathemat-ical objects and humans in the state of nature using this phrasing of "consider as." Laird connects "con-sidering" a conception with getting at the "nature" of it, what Laird identifies as a universal, but he fails to distinguish that not all cases of considering a conception will involve this (1968, 148). Indeed, most human reasoners—most of whom lack the combination of leisure, method, and sufficient curiosity—simply "consider" conceptions in ways that aid in satisfying their mundane passions (i.e., they do so in the service of prudence for this or that end). My aim is to distinguish mundane cases of considering, like Hobbes describes in *Leviathan* 4 (2012, 58–60; 1651, 15–16), from considering as that is driven by curiosity in the pursuit of *scientia*. Leibniz recognized the significance of 'considering as' in Hobbes's understanding of mathematical objects and universals and underlined *considerare* in an edition of *De corpore* owned by Boineburg (see Goldenbaum 2008, 74). Jesseph links Hobbes's view of bodies considered as points to Leibniz's account of "incomparably small" parts of lines or surfaces (2008, 226). Nauta (2012) draws attention to Leibniz's note, made in an introduction to Mario Nizolio's work, that a view similar to Nizolio's, in which one considers aspects of a thing while ignoring others, "has largely been inculcated in many by Thomas Hobbes" (Nauta 2012, 52). Describing Hobbes's use of 'consider as' in the account of 'person' (2012, 244; 1651, 80; but cf. change in 1668, 79), Brito Vieira asserts that the "technical sense" of 'to consider' is " 'to understand as in the view or sense of the law; to construe, to hold legally' (*OED*, s.v., IV.16)" (2009, 147; see also Brito Vieira 2021, 188). However, *pace* Brito Vieira, this quotation from the *OED* that Brito Vieira provides is not from the entry for 'consider' but rather the entry for 'intend' (cf. also the definition related to the law for 'intendment', *OED*, s.v. 4). While the *OED* entry for 'consider' does not include the "technical" usage claimed by Brito Vieira, it notes a usage aligned with a general cognitive ability to focus on this or that *aspect* of something: "To regard in a certain light or aspect; to look upon (*as*), think (*to be*), take for" (*OED*, s.v. 'consider', 10a). This more general understanding of 'consider' does not exclude legal ways of considering as (e.g., one could be considered as the owner of some property for a certain purpose), and it makes sense of Hobbes's usages outside of legal contexts, such as with regard to mathematical objects.

[17] Abizadeh argues that conceptions for Hobbes are "intrinsically aspectual" and "inherently com-parative" and that these features explain how "creatures incapable of having either universal or abstract conceptions could possibly invent universal names" (2017, 21; 2018, 81ff). My account of how we ar-rive at the universals of prudence, which are universals of "lesse extent" (cf. 2012, 52; 1651, 13) such as

SCIENTIFIC KNOWLEDGE FROM MAKING AND THE MECHANICAL MIND   59

The starting point for considering in this way is a large group of conceptions received in sense over time, and the stopping point of prudential categorization is individual human desire: if I desire to avoid the loss of a food source, then I judge similarities among my many conceptions that I take to be related to *that* consequent. I stop once my categorization is sufficient to satisfy my desire. Most humans will be content with this and go no further, because continuing on to gain *scientia* "serves them to little use in common life" (2012, 74; 1651, 21).

In contrast to the regulated trains of imaginations responsible for prudence, which seek to discover some antecedent that has often or always preceded a desired consequent, the regulated trains responsible for *scientia* result when "imagining any thing whatsoever, wee seek all the possible effects, that can by it be produced" (2012, 40; 1651, 9). To find such *productions*, one must make something. Indeed, Hobbes emphasizes in *Leviathan* 5 that this involves not predicting some consequent but rather the knower learning "what effects he could do with it" (2012, 68; 1651, 20) so that "when we see how any thing comes about, upon what causes, and by what manner; when the like causes come into our power, wee see how to make it produce the like effects" (2012, 72; 1651, 21). Some of these explorations involved in seeking *scientia* may produce little or no immediate payoff, but the passion curiosity—what Hobbes identifies as the "*Desire* to know [1668: *sciendi*] why, and how"—drives without regard to potentially wasted efforts.[18] The few humans that do undertake the work to gain *scientia* are driven by this passion of curiosity and have the leisure to so, as well as a method.[19] Doing so requires them to use the most

'nocturnal predator', agrees with Abizadeh. However, *pace* Abizadeh, my account denies that the pursuit of *scientia* is "inherently comparative." Rather than all being inherently comparative, which Abizadeh explains as "enabl[ing] the formation of universal names (for example, on the basis of the similarities and differences we conceive in sundry intentional objects)" (2017, 21; 2018, 81ff), Hobbes is explicit that the universals of *scientia*, such as the necessary proposition that 'all triangles are such that their angles sum to two right angles', are discovered only on the basis of a "consequence found in *one particular*" (2012, 54; 1651, 14; emphasis added). While inferences to the universals of prudence are from *many* conceptions, the inferences to the universals of *scientia* are from one, specially-considered-as conception. Pettit employs a strategy similar to Abizadeh's by proposing that Hobbesian minds are subliminally sensitive to similarities and dissimilarities in nature and that language allows humans to reason concerning those shared qualities (2008, 36). However, this likewise fails to account for Hobbes's view that the universals of *scientia* are not discovered by comparing similarities or dissimilarities but by a discovery made in a single conception.

[18]   Hobbes states in *Elements* that because non-human animals lack curiosity when a "beast seeth anything new or strange to him, he considereth it so far only as to discern whether it be likely to serve his turn, or hurt him" (EL I.IX.18).

[19]   I suggested in chapter 2 that in Latin *Leviathan* 6 Hobbes differentiates curiosity from other passions by using *scire* and *scientia* to describe the form of knowledge it seeks (knowledge of actual causes) and *cognoscere* to describe what humans and other animals share in seeking antecedents of desired effects (2012, 86–87; 1651, 26; 1668, 28; see fn. 30 in chapter 2). Tabb (2014) rightly notes that Hobbes's definition of 'curiosity' in *Elements* as "appetite of knowledge" (EL I.IX.18) does not distinguish the type of knowledge for which that passion is responsible. Tabb suggests that of the two options in *Elements*—"sense, or knowledge original" or "science or knowledge of the truth of propositions" (EL I.VI.1)—Hobbes would have had the former in mind because he "does not distinguish [curiosity] on linguistic grounds" (Tabb 2014, 21–24, here 22). As a result, Tabb argues that curiosity is unique as a passion because it is not directed toward some particular effect directly related to one's safety or welfare.

60 HOBBES'S TWO SCIENCES

abstract of names, such as 'body' and 'motion', and allows them to discover necessary propositions, such as "Body is extended," "Every man is an animal" (Hobbes 1981, 239; OL I.34; EW I.38), or "A rectilinear triangle has three angles equal to two right angles" (Hobbes 1981, 243; OL I.36; EW I.40). The remainder of this section will discuss this cognitive act that I am referring to with 'considering as' before turning in the next section to how it is used to reach a conceptual foundation.

In *Anti-White* II.1, Hobbes suggests that there is more than one way to take apart a conception by considering this or that feature of it. Before criticizing Thomas White's claim that the world is finite, Hobbes asserts that he must first answer "what it is to divide and to compound" (1973, 108; 1976, 29). Hobbes offers two senses of considering a conception: first, one can divide a conception by ignoring a non-essential accident that is "inside" the image of it; and second, one can consider a conception by paying special attention only to *necessary* conceptual constituents, which Hobbes identifies as part of its "nature." Both ways are types of considering, but the former is a case of "dividing" while the latter is what I am identifying as 'considering as'. Hobbes makes the distinction as follows:

> We are said to divide [*dividere*] something when we consider first the thing and then something smaller contained in it. For example, when I am considering a man as a rational animal [*considere hominem ut animal rationale*] I am not dividing a man; but after I have considered the same man as being composed of a head, shoulders, arms, etc., then I am said to have divided him. (1973, 108; 1976, 29)

As an example of dividing, I might call up the image of a pizza (PIZZA), mentally divide it into eight slices (EIGHT + EQUAL + SLICE), and then consider only

---

While Tabb rightly notes the role language (i.e., propositions) plays according to Hobbes's account in *Elements* (EL I.VI.1; see also *Leviathan* 5, 2012, 72; 1651, 21), *Elements* does not imply that science is *exclusively* reasoning in language. Indeed, several articles later Hobbes more completely defines science as "evidence of truth, from some beginning or principle of sense" (EL I.VI.4). This appeal to 'evidence', which I have argued elsewhere is achieved by examining one's *conceptions* from sense and finding relations among them (see Adams 2014a, 413–414 for discussion of 'evidence' in EL I.VI.3), makes clear that conceptions are the "first principle" of *scientia* and thus constructions from those conceptions, which may be remembered with names and propositions, can be driven by curiosity. Rather than connecting curiosity with prudence, i.e., in the language of *Elements*, "sense, or knowledge original," as Tabb (2014, 22) does, Hobbes suggests in *Elements* that there are two "degrees of curiosity" such that this same passion is responsible for the invention of names and the supposition of causes of what "might produce" some effect (EL I.IX.18); from the latter "is derived all philosophy" including astronomy and natural philosophy (regarding these two "degrees," see also *Leviathan* 5, 2012, 68; 1651, 20; *Leviathan* 11 claims natural religion arises from it as well [2012, 160; 1651, 51]). Furthermore, Hobbes excludes "original knowledge" and prudence, which is "but a Memory of successions of events in times past," from the definition of 'philosophy' in *Leviathan* 46 (2012, 1052; 1651, 367). According to my account, curiosity drives the search for *scientia*, but few humans possess *scientia* because it also requires leisure and method. Hobbes's definition of 'philosophy' in *De corpore* I.2 (Hobbes 1981, 175; OL I.2; EW I.3) suggests that its two paths are either driven by curiosity in geometry and civil philosophy (actual causal knowledge provides knowledge of effects) or, as I argue in chapter 5, transformed by curiosity in natural philosophy when knowers posit possible causes by borrowing them from geometry.

SCIENTIFIC KNOWLEDGE FROM MAKING AND THE MECHANICAL MIND   61

a SLICE that represents 1/8 of the PIZZA. Dividing PIZZA in this way requires ignoring non-essential accidents (the original shape and magnitude), but the remaining consideration is still the same type of thing. The SLICE considered as 1/8 of the original is still PIZZA, just smaller and differently shaped. Considering by dividing occurs when in prudence I categorize this or that conception by picking out some non-essential feature made salient by *my* desires. The conception signified by 'coyote' can remain in my mind whether I consider it 'yellow furred' or not, but were I to ignore an accident like 'extended', I would no longer be considering COYOTE but have *nothing* at all in the mind since "no body can be conceived without extension, or figure" (OL I.92; EW I.104).[20]

An example of dividing as a mode of considering in *De corpore* VII.5 makes clear that the mental act of considering happens with conceptions, not names: "[. . .] to *make parts*, or to *part* or divide space or *time*, is nothing else but to consider one and another within the same [*quam in ipso aliud atque aliud considerare*]; so that if any man *divide* space or time, the diverse conceptions [*conceptus*] he has are more, by one, than the parts he makes [. . .]" (OL I.85; EW I.95–96; emphasis original).[21] Hobbes further emphasizes that considering is a *mental* act stating, "I do not mean the severing or pulling asunder of one space or time from another [. . .] but diversity of consideration," and affirms that this is "a work [*opus*] not of the hands, but of the mind" (OL I.85; EW I.96). Mental composition is also possible: "[. . .] as if one should reckon first the head, the feet, the arms, and the body, severall, and then for the account of them all together put man" (OL I.86; EW I.97). Imagining conceptions of these disparate parts of a human body all at once results in imagining the conglomerate conception HUMAN.

Later in *De corpore* VI.2, Hobbes contrasts these two modes of considering— dividing and considering as—by differentiating considering a man's "bulk" from considering his "nature":

> But by 'parts' I understand in this place not the parts of the thing itself but the parts of its nature [*naturae*], so that by parts of a man, I do not understand head, shoulders, and so on, but figure, quantity, motion, sensation, reasoning, and similar things, which are accidents which assembled at the same time constitute the whole man—not his bulk but his nature. (Hobbes 1981, 291; OL I.60; EW I.67)[22]

---

[20] Leijenhorst sees a version of the primary/secondary quality distinction in Hobbes's discussion of *accidentia propria* and *accidentia communia*, suggesting that magnitude and motion would be *accidentia communia* (2002, 160). This plausibly holds for 'magnitude', but Hobbes explicitly lists "to be at rest" and "to be moved" as proper accidents alongside color and hardness (OL I.93; EW I.104). Slowik (2014, 78) suggests understanding motion/rest as a "disjunctive primary property."

[21] Hobbes defines 'one' in *De corpore* VII.6 as "when space or time is considered [*consideratur*] among other spaces or times" (OL I.85; EW I.96; cf. MS 5297, Hobbes 1973, 450–451).

[22] See also *De corpore* VII.8 (OL I.86; EW I.96–97).

62   HOBBES'S TWO SCIENCES

The former are *divided* parts, such as considering just a HEAD or SHOULDER of a MAN, but the latter are from considering as: these accidents are the conceptions that arise as MAN is received in sense (BODY + ANIMATE + RATIONAL). One *considers* MAN *as* by attending to the conception RATIONAL and not focusing one's attention upon other conceptions like BODY. However, although one can consider one of these in isolation from the other, unlike SHOULDER, one cannot remove one of *these* accidents and still be considering MAN. To imagine a shoulder-less man is still to imagine MAN, just like imagining SLICE is still imagining PIZZA, but one cannot consider as MAN without BODY. As will be discussed below, this necessity, evident to a knower by considering as, will be remembered in a necessary proposition: 'Every man is an animal' (Hobbes 1981, 239; OL I.34; EW I.38).

This ability a human mind has to "consider as" conceptions anticipates Berkeley's recognition that

> [...] it must be acknowledged that a man may consider a figure merely as triangular, without attending to the particular qualities of the angles or relations of the sides. So far he may abstract, but this will never prove that he can frame an abstract general inconsistent idea of a triangle. In like manner we may consider Peter insofar as he is a man or insofar as he is an animal [...] (PHK, Intro. §16; *Works* II.35)

Like Berkeley, Hobbes allows an abstractive process from a particular conception that focuses attention on some necessary property of a conception, such as having three sides. As mentioned in chapter 2, like Berkeley, Hobbes denies that such an abstraction can produce either wholly abstract ideas or is a guide to learning about independently existing essences: Hobbes asserts that abstract names are "severed (not from Matter but) from the account of Matter" (2012, 58; 1651, 16).[23] Many of Hobbes's criticisms of Scholastic philosophy are directed at what he saw as such

---

[23] Brandt articulates the similar position in which Hobbes and Berkeley find themselves: both understand ideas/conceptions as particular while holding that humans can nevertheless reason universally (Brandt 1928, 236–239). Brandt contrasts Hobbes's account of universals with Berkeley's view that an idea of a particular line, which is "with regard to its signification general," can be used in such a way that it "represents all particular lines whatsoever" (1928, 237). Brandt recognizes Berkeley's solution—one idea represents "all other particular ideas of the same sort"—but Brandt's account of Hobbes's understanding of universals is lacking. Rather than explaining how Hobbes thought humans *discover* universality for *scientia* (e.g., the discovery that 'All triangles are such that their angles sum to two right angles' from a single particular), Brandt changes the subject to what happens when a universal name is *used*. He argues that Hobbes held that "we conceive the universal" by names so that "psychologically when we think universally [...] a succession of concrete ideas crops up" (1928, 232). In other words, while Brandt treats Berkeley's solution for *arriving* at universality (the particular idea represents all as a sign), the treatment of Hobbes that he provides only concerns what happens in the mind when a name like 'triangle' is used, which could only occur *after* universality has already been achieved. Against Brandt's claim that Berkeley's "solution" was not shared by Hobbes, Hobbes argues against Thomas White in *Anti-White* IV.2 that "the imaginary surface caused by the sight of the running water was an image of the water, not as water [*ut aquae*], but as body [*ut corporis*], and that [ii] the image therefore represents [*repraesentare*] not that particular water but any water, or air, or body of the same size and shape" (1973, 126–127; 1976, 53). I discuss this example below in more detail in section 3.5 (and have treated it within

SCIENTIFIC KNOWLEDGE FROM MAKING AND THE MECHANICAL MIND   63

attempts (see Foisneau 2021). Furthermore, beyond this ability to "consider as", such as considering a figure as triangle, which Berkeley allows, the Berkeleian account, like Hobbes's, accommodates the mind's ability to imagine by dividing or compounding non-necessary properties like when one imagines "a man with two heads or the upper parts of a man joined to the body of a horse" or even when one "consider[s] the hand, the eye, the nose, each it self abstracted or separated from the rest of the body" (PHK Intro. §10; *Works* II.29).

In the act of considering as, the Hobbesian mind never leaves the image of the particular entirely behind (discussed in the next section). Indeed, in *Anti-White* XXX.36 Hobbes indicts much of philosophizing as being a game of words with little connection to particular conceptions received in sense: "[...] the speech of the greatest number of those who philosophize is accompanied by no thoughts of things, but words received rashly, variously mingled, and compounded into propositions, until finally they seem to signify something excellent" (1973, 366; 1976, 388).[24] Hobbes accuses Thomas White of having fallen astray in this manner when "discussing abstractions [*abstractiones*] from place, from continuous parts, from time, and the like" (1973, 366; 1976, 388). Hobbes's alternative is "speaking clearly from a consideration [*considerationem*] of place, parts, and time" (1973, 366; 1976, 388).

Against the abstractionist view, Hobbes declares that it is "not difficult to consider [*considerare*] place apart from a consideration [*consideratione*] of body" since any unphilosophical person (*rusticus*) understands PLACE as simply SPACE in which a body can be. An unphilosophical person would think a speaker "mad" (*insanire*) who claimed that place is abstract (*abstrahere*) (Hobbes 1973, 366; 1976, 388). Hobbes likewise claims that one cannot think of TIME by wholly abstracting it from particulars. Berkeley later echoes this appeal to unphilosophical intuitions as a reason for rejecting wholly abstract ideas: "The generality of men which are simple and illiterate never pretend to abstract notions" (PHK Intro. §10).

Hobbes's criticisms of the Cartesian attempt to show that mind can exist independently of body are founded on this same complaint (e.g., Hobbes's Second Objection; CSM II.122; AT VII.172–173). As should be clear, Hobbes is sanguine

---

the context of Hobbes's criticisms of Descartes; see Adams 2014a, 417–20), but Brandt's attempted distinction between Berkeley and Hobbes, which claims that Hobbes "gets over the problem [of achieving universality] by jumping from ideas to names" while Berkeley "chooses another way" (Brandt 1928, 237), does not hold water. Although Hobbes does not use the Berkeleyan terminology that a particular idea is a sign for all, Hobbes does claim that the mental act of *considering* a particular *as*, e.g., considering an image of water as an imaginary surface, can be taken to *represent* all surfaces similar to it. As I will argue below, the Hobbesian emphasis on language in the context of discovery is simply that without names what is discovered would be forgotten.

[24] I have modified Jones's translation here and elsewhere. See also *Leviathan* 4 (2012, 58; 1651, 16) and *De corpore* III.3–4 (Hobbes 1981, 227–231; OL I.28–30; EW I.31–34).

64   HOBBES'S TWO SCIENCES

to considering 'thinking' apart from 'body', but "abstraction' [*abstrahere*] or 'exist-ence apart from them'" is an "abuse" (Hobbes 1981, 231; OL I.30; EW I.33):

> The gross errors of certain metaphysicians take their origin from this; for from the fact that it is possible to consider thinking without considering body, they want to infer that there is no need for a thinking body [. . .] These meaningless vocal sounds, "abstract substances," "separated essences," and other similar ones, spring from the same fountain. (Hobbes 1981, 231; OL I.30; EW I.34)

Clearly, then, not all abstraction is in error; indeed, Hobbes distinguishes his own geometry as "abstract" when compared to natural philosophy (OL I.314; EW I.386). Abstraction rightly understood must always be tethered to conceptions of particulars.

Hobbes's claims about the impossibility of wholly abstract conceptions did not prevent him from seeing benefit derived from the act of considering as. In the next section, I argue that removing the "unnecessary parts" (Hobbes 1655, OL I un-numbered) from particular conceptions via considering as allows a finite human to reach a foundation for a construction that will follow. From such a foundation, they can make something like a SQUARE in geometry or PEACE in civil philos-ophy and use language to remember their discovery.

3.3 Considering As to Reach a Foundation for Making

The ability to consider as requires effort by a human knower, in contrast to the pas-sive "decay" that with accumulated sense experiences and comparative judgments of similitudes produces associations and eventually prudence. But how does this ability to consider as work? An example—not Hobbes's—will motivate further discussion of Hobbes's understanding of this mental act. I may consider my con-glomerate conception of a bouncing ball (RED + BOUNCING + BALL) in front of me as a child's play toy. Considered in this way, I may imagine whether my niece would enjoy playing with it, given her likes and dislikes. Alternatively, I may con-sider that same conception as a BODY by ignoring its apparent color, motion, and other properties. Such considering as could lead me to wonder what would result from an impact with another ball in motion (considered as a BODY + MOTION). Finally, I may consider that conception as a geometric sphere (SPHERE) and wonder about its properties, or even consider part of that SPHERE as a LINE after tracing its diameter. How I consider this unified, conglomerate conception RED + BOUNCING + BALL depends upon my interests and aims. If buying a gift for my niece, I consider it in the first way; such considering (putting into an account) would result from prudence and my desire to bring about my niece's pleasure. However, if I am driven by curiosity and have leisure, I may consider it as

SCIENTIFIC KNOWLEDGE FROM MAKING AND THE MECHANICAL MIND   65

a geometrical object in the final way; such considering as driven by curiosity would seek to discover all possible effects from LINE, and could provide a foundation for discovery and, ultimately, after construction, maker's knowledge of SPHERE.

Say that do I follow that final way: I consider my conglomerate conception RED + BOUNCING + BALL as SPHERE, my conception of the apparent path the ball takes in motion while bouncing as STRAIGHT + LINE, and the floor on which it bounces as a FLAT + SURFACE. The conception received in sense is not of such idealized geometrical objects but rather of a rubber ball that appears red to me and has many surface imperfections from bouncing on rough surfaces and of a slightly sloped, uneven floor surface. Nevertheless, say that driven by curiosity I ignore these features and consider only SPHERE, STRAIGHT + LINE, and FLAT + SURFACE. Next say that I abstract further, considering my conception only as STRAIGHT + LINE. Abstracting further I may consider STRAIGHT + LINE as having been made by the motion of a body considered as if it were an instantaneous POINT. Finally, imagine that I consider only POINT.

The mind creates an artificial mental object at each step. These artificial objects are *drawn from* conceptions received in experience, but they are not found anywhere in nature as such. As with 'thought' in his criticisms of Descartes, Hobbes explicitly denies independent existence to artificial geometrical objects like these, stating that there are no such things as breadthless lines in the world, since a line is only "a body whose length is considered without its breadth" (EW VII.202). Just like the distinction between memory and imagination results from "the considerations of the sentient [*considerationum sentientis*]" and not from "things themselves" (OL I.325–326; EW I.398), artificial geometrical bodies, like points and lines, result from considerations of human knowers. Just like sculptors take a hunk of rock and carve away the "unecessary parts," so too the human knower carves away the parts that are unnecessary given their interests and aims. Importantly, denying independent existence does not prevent these abstractions from being treated in the mind *as if* they were independent when they serve as a foundation for the construction of something entirely novel (what I describe below in section 3.4 as Making Type 2).

This act of considering as the red ball occurs by activities that Hobbes calls analysis and synthesis. There has been much discussion of Hobbes's understanding of and source for analysis and synthesis, but in practice he shows that he understands these activities fairly simplistically.[25] Indeed, he frequently uses phrases like "subtracting and adding" or "resolving and compounding", respectively, as synonyms for these acts of the mind. Hobbes's simple point is this: the mind has the ability to take apart its conglomerate conceptions and put them back together, or it can put together new ones from disassembled ones. In *De corpore* I.3, Hobbes

---

[25] For example, see Jesseph (1999, 224ff) and Talaska (1988). Hintikka and Remes (1974) and Otte and Panza (1997) treat the history of analysis and synthesis.

66   HOBBES'S TWO SCIENCES

calls the process of adding and subtracting "computation [*computationem*]" and claims "[w]hatever we add or subtract, that is, subject to reason, we are said to consider [*considerare*]" (Hobbes 1981, 181; OL I.5; EW I.5).[26] As argued in chapter 2, knowers *consider* conceptions (often using names that signify judgments concerning similitudes in them) by adding to or subtracting from them in reasoning syllogistically, but now we can add to this account: Knowers *consider as* when, driven by curiosity, they focus on parts of the nature of a conglomerate conception and ignore other parts of it.

I will now examine an example of considering as that Hobbes uses in *De corpore* VI.6. The remainder of this section treats the first stage (arriving at an abstracted foundation for construction through resolution), and the next section discusses the second stage (constructing from that foundation to gain causal knowledge through composition). In each step of resolving some feature or other is excluded, not by "dividing" non-essential accidents like particular size or shape, but by considering as:

> [...] let any conception or idea of a singular thing [*conceptu sive idea rei singularis*] be proposed, say a square. The square is resolved into: plane, bounded by a certain number of lines equal to one another, and right angles. Therefore we have these universals or components of every material thing: line, plane (in which a surface is contained), being bounded, angle, rectitude, and equality. If anyone finds the causes of these, he will put them together as the cause of the square. (Hobbes 1981, 293; OL I.61; EW I.69)

This example begins middle of the way through a resolution since its starting point is *already* a geometric figure—a *particular* SQUARE. As mentioned already, Hobbes holds that geometric figures like squares do not exist *qua* squares in the world. Squares are made of lines that likewise do not exist *qua* lines in the external world; instead, lines are from bodies with breadth that humans "consider as" having no breadth (EW VII.202).[27] A line is "made by the motion of a point"

---

[26] Many things beyond names can be subjected to computation. The account of computation in *De corpore* I.3 immediately follows the example discussed in chapter 2 where a perceiver receives BODY, ANIMATE, and RATIONAL via the senses of sight and hearing (Hobbes 1981, 177–179; OL I.3–4; EW I.4). Hobbes claims he has "shown well enough with these examples how the internal reasoning of the mind works without words [*sine vocibus*]" (Hobbes 1981, 179; OL I.4; EW I.5), suggesting that computation is also used with magnitudes, bodies, motions, conceptions, and names (Hobbes 1981, 179–181; OL I. 4–5; EW I.5). He distinguishes three types of analysis in *De corpore* XX.6—analysis according to indivisibles, powers of lines, and motions—but I do not focus on this distinction since my interest in analysis and synthesis is on commonalities between first philosophy, mathematics, and civil philosophy.

[27] See also *Anti-White* II.8 where Hobbes criticizes White's understanding of measurement and discusses three ways of measuring that relate to 'length' (*longitudo*), 'breadth' (*latitudo*), and 'thickness' (*crassities*). Regarding 'length', he claims that it designates the motion of a body when its magnitude is not "considered" (*consideratur*) (1973, 114; 1976, 37–38). A body considered in this way is what in Greek and Latin has been identified as *stigmhn*/*kentron* and *punctum*, respectively, and the motion of a point of this sort is what forms a line. The second way of measuring involves a body's motion where

SCIENTIFIC KNOWLEDGE FROM MAKING AND THE MECHANICAL MIND   67

(Hobbes 1981, 297; OL I.63; EW I.70), and a point is a body considered not as that which "has no quantity, or which cannot by any means be divided; for there is no such thing in nature" but rather as a body "whose quantity is not at all considered, that is, whereof neither quantity nor any part is computed in demonstration" (OL I.177; EW I.206).[28] As confirmation of this understanding of geometrical objects, Hobbes appeals to Proclus on Apollonius's understanding of the definition of line, citing Henry Savile's reference: "See this of Proclus cited by Sir Henry Savile, where you shall find the very word *consider*" (EW VII.202; emphasis original).[29]

Were Hobbes beginning this resolution from a unified, conglomerate conception received in sense, he might have instead suggested beginning from a conception like that of a square closet door.[30] From that conglomerate conception, the knower would subtract particular conceptions from each of the senses, such as apparent color, scent, texture, and so on, by ignoring them. In other words, the knower would begin *considering* CLOSET DOOR *as* a SQUARE by not attending to those other particular conceptions that are received in sense, thus creating an artificial geometrical object. The conception CLOSET DOOR is not of a square as such, but the knower with leisure and curiosity could use this method of resolving to consider it *as* a SQUARE.

The first stage of the resolution of SQUARE that Hobbes actually provides considers it as composed of "plane, bounded by a certain number of lines equal to

its length is considered without breadth (*sine latitudine*), which is to say involves measurement of a line, which Hobbes connects with *grammhn* (i.e., γραμμήν). In *Anti-White* Hobbes is careful to distinguish the actual bodies being measured and the mathematical objects created by considering those actual bodies as without magnitude or breadth, saying about a 'line' that it is "not without breadth [*sine latitudine*], but is considered [*consideratur*]" as such (1973, 115; 1976, 38).

[28] Sometimes Hobbes does not criticize Euclid himself but how he has been interpreted. In *Six Lessons* (Lesson I) Hobbes argues against the common understanding of Euclid's definitions and, while he admits that a straightforward translation of Euclid's definition of 'point' is "that of which there is no part" (EW VII.200), he claims that this definition is "neither useful nor true" for those who cannot "interpret candidly [. . .] nor accurately." To understand "no part" properly, an "accurate interpreter" must distinguish between "undivided" and "indivisible" (EW VII.201). Whereas the latter has been taken by Euclid's interpreters to be his meaning, Hobbes asserts the former is the only way to make sense of the definition since "division is an act of the understanding," and, unsurprisingly, the proper translation on this account accords with Hobbes's own view that "a point is that body whose quantity is not being considered" (EW VII.201). Immediately following this revised interpretation of Euclid, Hobbes directs the reader to his definition of 'consider' in *De corpore* I.3: "And *considered* is that, as I have defined it in chap. I at the end of the third article, which is not to put to account in demonstration." Considering a body in this way, then, means that one does not "put to account" the parts of a body when using it as a point within a geometrical demonstration (OL I.5; EW I.5).

[29] Savile's report of Proclus on Apollonius, and the mention of 'consider', can be found in *Praelectiones Tresdecim in Principium Elementorum Euclidis* (Savile 1621, 65).

[30] That geometrical objects arise from considering conceptions caused by natural bodies provides the first step to understanding why Hobbes thinks mathematics can be used in natural philosophy. Instead of holding that Hobbes simply assumed "that nature had an underlying mathematical structure which was not apparent to sense," as Peters (1956, 56) does, I argue in chapter 5 that Hobbes thought natural philosophy should ideally proceed by humans treating natural bodies *as if* they were mathematical bodies and then principles from geometry could be borrowed within natural-philosophical explanations.

68   HOBBES'S TWO SCIENCES

one another, and right angles" (Hobbes 1981, 293; OL I.61; EW I.69). How does one do this? To resolve SQUARE into these conceptions one must focus on a feature of SQUARE at a time and ignore others. Language is not yet required. Indeed, the example is clear that this work of resolving is done by examining a singular conception (*conceptu sive idea rei singularis*), not a name or proposition. Furthermore, in Hobbes's second example of resolving immediately following the SQUARE example, GOLD is resolved not into names but into "the ideas of being solid, visible, and heavy" (Hobbes 1981, 293; OL I.61; EW I.69).[31] How is the resolution so far of a SQUARE not a case of mere "dividing," similar to considering a conception MAN by dividing it into HEAD, SHOULDERS, ARMS, and so on? In contrast to the resolution of SQUARE so far, dividing SQUARE could involve considering it with a smaller magnitude; the result would still be a SQUARE. Instead, resolving thus far involves attending to the necessary accidents of the conception of the SQUARE, evident from the fact that if one of them were removed, e.g., RIGHT ANGLES, then the resulting conception would no longer be a SQUARE.

Resolving continues into "line, plane (in which a surface is contained), being bounded, angle, rectitude, and equality." What is the difference between this second stage and the first stage? The first stage offers BOUNDED BY A CERTAIN NUMBER OF LINES EQUAL TO ONE ANOTHER, but the second stage splits this conglomerate into components by considering it as BEING BOUNDED, LINE, and EQUALITY.[32] Hobbes indicates that there are "causes of these" but does not provide them here; however, given what Hobbes says elsewhere, we can determine what he means. The causes of *a* LINE according to Hobbes should be obvious; as noted already, we imagine a LINE when moving a POINT (Hobbes 1981, 297; OL I.63; EW I.70), and so a particular conception LINE resolves into BODY and MOTION. Thus, in the mind there would be the conception of *a* BODY, i.e., a conception of a mind-independent, extended thing (OL I.91; EW I.102), and a particular instance of MOTION, i.e., "the idea of a body crossing, now through this, now through that space in continual succession" (OL I.83; EW I.94). Likewise, BEING BOUNDED resolves into PLACE, since PLACE is a part of imaginary SPACE coincident with the magnitude of some body (OL I.93; EW I.105).[33]

How would EQUALITY and RECTITUDE resolve? Both of these relate to the *type* of lines needed to make SQUARE. Knowing how to make a line by moving a point does not distinguish between straight and crooked lines or between lines of different lengths. Thus, we resolve the particular instances of LINE that are part

---

[31] I do not discuss GOLD because Hobbes provides fewer details concerning the stages of its resolution.

[32] Conceptions like EQUALITY and MOTION will *eventually* be marked or signified with abstract names (see Hobbes 1981, 227–229; OL I.28–29; EW I.31–33), but Hobbes's examples of SQUARE and GOLD are clear that names are not needed to do the work of resolution described so far.

[33] For discussion of Hobbes on imaginary space, see Jesseph (2015; 2016, 136) and Slowik (2014).

SCIENTIFIC KNOWLEDGE FROM MAKING AND THE MECHANICAL MIND   69

of a SQUARE when we draw or imagine each of the four lines as straight, understood as the shortest line drawn between two points "whose extreme points cannot be drawn farther asunder without altering the quantity, that is, without altering the proportion of that line to any other line given" (OL I.153–154; EW I.176). Similarly, EQUALITY resolves into sameness in length of four lines. Lastly, we resolve SQUARE by determining that these four lines must be connected with right angles, each understood as an angle "whose quantity is the fourth part of the perimeter" of a circle (OL I.161–162; EW I.186–187). Thus, we first make a circle, whether in our imagination or on paper, determine the "quantity" of the angle made by one-fourth of the perimeter, and then compare this to the angles of the SQUARE. Given his view that all conceptions are images, Hobbes equates moving a point on a piece of paper with the *conception* of a moving point forming a line.

I have argued elsewhere that the "causes of these" to which Hobbes alludes in this example are what he calls the "simplest conceptions" in *De corpore* VI.6 (Adams 2019; 2014b). Examples of these simples mentioned already are BODY, PLACE, and MOTION.[34] Hobbes describes these as "universals," insofar as they are part of the composition of all conglomerate conceptions received in sense, and, as simples that cannot be further resolved, they serve as the terminating points of the resolution of a conception of a body (like CLOSET DOOR) being considered as a geometric figure like a SQUARE. Hobbes holds that all knowers have *awareness* of these "simplest conceptions" even without having a definition of them in language, and he places *conceiving* of simplest conceptions like PLACE before having definitions of them in language: "For whoever correctly conceives (*concipit*) what a place is [...] must know the definition: that a place is a space which is completely filled or occupied by a body" (Hobbes 1981, 295; OL I.62; EW I.70).

Anyone who has received any conception whatsoever in sense, and thus possesses BODY, will *ipso facto* have PLACE in the mind, since the simplest conceptions are "contained [*contineantur*] in the nature of singular things" (Hobbes 1981, 293; OL I.61; EW I.68–69). As a result, the definition (in language) for 'place' is merely an "explication" of that already-held conception (Hobbes 1981, 295; OL I.62; EW I.70; I discuss definition by explication in chapter 4). Although anyone with experience has these conceptions, Hobbes does not hold that everyone *correctly* conceives BODY or PLACE; indeed, *mis*conception is possible.[35] Hobbes describes awareness of these simplest conceptions as "apprehending," differentiating apprehending (*cognoscere*) the simplest conceptions from knowing the causes of things (*scire*) by "putting together": "[...] before the causes of those

---

[34] Since not all bodies are moving, not all conceptions will contain MOTION. Slowik has suggested understanding MOTION as disjunctive, such that all conceptions contain either MOTION or REST (2014).

[35] For example, he argues in *Anti-White* III.5 that Aristotle's definition of 'place' is not true (Hobbes 1973, 120; 1976, 44–45).

70   HOBBES'S TWO SCIENCES

[singular] things can be known [*sciri*] it is necessary to know [*cognoscere*] which things are universals" (OL I.61; EW I.68).[36]

The mind works with particular conceptions from the beginning to the end of this resolution in *De corpore* VI.4. One has SQUARE in mind when imagining a particular square; Hobbes is clear that the resolution is of a singular thing (*rei singularis*). After the example, he declares that "by this same method of resolving things into other things one will know what those things are, of which, when their causes are known and composed one by one, the causes of all singular things [*rerum singularium*] are known" (Hobbes 1981, 293; OL I.61; EW I.69). Each part of a particular conception is also a particular conception (so resolving the particular conception SQUARE yields the particular conception LINE), which aligns with Hobbes's claims elsewhere that only *names* are universals (e.g., EL I.5.6; 2012, 52; 1651, 13) and that it is an error to think that "the idea of some thing is a universal, as if there might be in the mind of some man an image of some man which is not that of any one man, but of man *simpliciter*" (Hobbes 1981, 277–279; OL I.53–54; EW I.60). Thus, it is unsurprising that in an earlier mention of SQUARE in *De corpore* I.3, Hobbes states that the necessary accidents of SQUARE are themselves each particular: "For the mind can conceive a quadrilateral figure without the conception of a rectangle, and it can conjoin these singular conceptions [*concepta haec singula*] into one conception or one idea of a square" (Hobbes 1981, 179; OL I.4; EW I.4).

Understanding the resolution of SQUARE as an act of the mind performed apart from language may seem strange given how language helps humans develop prudence. Recall the earlier discussion in chapter 2 of how the "internal reasoning of the mind works without words" (Hobbes 1981, 179; OL I.4; EW I.5) and how error is possible apart from language (*De corpore* V.1 Hobbes 1981, 271–272; OL I.50; EW I.56; see also *Leviathan* 4, 2012, 54; 1651, 15). For the case of *cognitio*, Hobbes's separation of the composition of individual conceptions in sense from the judgments *among* conceptions does not reflect how sense actually works. As Hobbes admits, sense in the actual world, by those who possess a stock of conceptions, is inherently comparative and includes judgments among conceptions (see EW I.393; Abizadeh 2017; 2018, 81ff; Barnouw 1980a, 122–123). Likewise, actual *prudential* reasoning by humans with a baseline of experiences involves the examining of conceptions alongside reasoning with language, requiring one to be sure that words used are not absurdities like 'incorporeal substance' or, in the phrasing of *Elements*, that one has "evidence" (see fn. 19 above).

My emphasis that resolving SQUARE involves conceptions alone—not language—reflects that Hobbes understands the act of discovery (*invenire*) needed for *scientia* to be essentially different from the development of prudence. Rather

---

[36] The Molesworth edition (EW I.68) and Martinich's translation (Hobbes 1981, 292–293) blur this distinction between *cognoscere* and *scire* by translating both with 'know'.

than accumulating over judgments of similitudes among many conceptions, *scientia* arises from consideration of a particular conception resolved down to its bedrock of particular conceptions, what Hobbes calls sifting to the bottom in civil philosophy (2012, 320; 1651, 107). We have reached the bottom of SQUARE, but resolving alone does not provide *scientia* since it does not provide causal knowledge, i.e., the knower has not yet *made* a SQUARE. This constructive act from the foundation reached is the subject of the next section.

3.4 Making as Discovery for *Scientia*

Composition, or synthesis, adds conceptions to a particular conception that serves as a foundation. I distinguish three types of Hobbesian making:

Making Type 1: putting back together a conglomerate conception beginning from its necessary parts reached through resolution;

Making Type 2: making something that is different from the original conception by beginning from a foundation reached in resolution; and

Making Type 3: when teaching (*docere*), skipping a resolution altogether, beginning with a thought experiment, and in synthesis constructing from that foundation.

I discuss the first two types of synthesis in this section and the third in chapter 4. What unites all three modes of synthesis is that they begin a construction from an abstracted foundation in which a knower has removed the "unnecessary parts," whether through analysis or by beginning with a thought experiment. From such a foundation one uses either private marks to remember the discoveries made, or public signs when teaching another. These two uses of language are so significant that Hobbes singles them out in *Leviathan* 4 as "speciall uses."[37] The first special use of language, discussed below in section 3.5, is when we "[. . .] Register, what by cogitation [1668: *cogitando*],[38] wee find to be the cause of anything, present or

---

[37] Although language in these "special" uses, i.e., *particular* uses, as Hobbes uses 'special' elsewhere (cf. 2012, 131 fn. d), has the same general function as language for other uses (i.e., marking our conceptions to aid in remembering and signifying for others), as Abizadeh notes (2015, 3 fn. 9), the *actions of the mind* that language marks/signifies in these uses are significantly different from others. Hobbes's use of 'art' shows he means this to be a special use of language such that it aids the development of philosophy, in particular geometry, natural philosophy, and civil philosophy (see discussion of these "arts" in *De corpore* I.7; Hobbes 1981, 183–185; OL I.6–7; EW I.7). See also *Six Lessons* on Hobbes's identification of 'art' with maker's knowledge (EW VII.183–184).

[38] I highlight the usage of *cogitando* in the 1668 edition because Hobbes uses this same term to describe non-linguistic activity in the triangle example in *De corpore* VI.11 (Hobbes 1981, 311; OL I.70; EW I.80). I discuss this example below, as well as the corresponding example from *Leviathan* 4 (2012, 52; 1651, 14) where Hobbes uses 'meditation' to describe this activity (for discussion of 'meditating' on one's conceptions, especially Hobbes's instruction to take instruction from one's own meditation rather than from the "authority of books" in *Leviathan* 4 [2012, 56; 1651, 15], see Adams [2014b, 50–51]).

72 HOBBES'S TWO SCIENCES

past; what we find things present or past may produce, or effect: which in summe, is acquiring of Arts" (2012, 50; 1651, 13). The second "special use" of language—showing others "that knowledge [1668: *scientiam*]" by teaching—is a focus of chapter 4.

Returning to the earlier example of the red bouncing ball will motivate discussion of the first two types of making. If a resolution ended by considering the red bouncing ball as a POINT, one could re-make the original conception by mentally re-assembling the components, i.e., a composition that rebuilt the original conglomerate conception RED + BOUNCING + BALL (Making Type 1). Alternatively, a synthetic process could begin from that same foundation—a POINT—and make some novel conception different from the original (Making Type 2). For example, I could add RECTITUDE and MOTION to POINT, either drawing on paper or imagining it, and thus make the conception STRAIGHT + LINE. According to Hobbes, I know the causes of this *new* conception since I myself produced it. Then I could add MOTION to STRAIGHT + LINE, imagining or drawing that line as moving in a single direction, to make SURFACE (Hobbes understands "a surface [*superficies*] [as made] from the motion of a line" [*De corpore* VI.6; Hobbes 1981, 297; OL I.63; EW I.70]). Although I began from the same starting point, constructing a SURFACE differs from re-assembly. As mentioned earlier, Hobbes states abstract names are "severed (not from Matter but) from the account of Matter" (2012, 58; 1651, 16), but cases of Making Type 2 show just how much "severing" Hobbes allows. Considerations are always tied to the conception of some bit of matter or other, but constructions may differ greatly from the original matter from which a resolution/composition began.[39] These abstractions are treated in the mind *as if* they were independent, though never apart from BODY, when they serve as a foundation for the construction of something entirely novel.

The remainder of this section will focus on re-assembling a SQUARE (Making Type 1). Recall that resolving SQUARE ended at termination points that Hobbes calls "our simplest conceptions." These are BODY, PLACE, and MOTION. These alone are not enough to put the square back together. In addition to a BODY considered as a POINT combined with MOTION, one must have *four* points and move them in a certain way to make four lines (four *straight* lines). Furthermore, one must make those four lines in a certain proportion to one another (equality understood as equal magnitude). Next, one must place these four lines together so that they form four right angles, which occurs by making a circle and marking one-fourth of the perimeter. Doing all of these actions in the correct order results in the

---

[39] Hobbes's development of geometry is thus an example of Making Type 2 making where one begins from simplest conceptions like BODY and MOTION and from them constructs new types of figures (e.g., a circle) and new types of motions (e.g., the "simple circular motion" of *De corpore* XXI). Talaska (1988, 214) has argued that Parts I–III of *De corpore* are in the form of a synthetic demonstration.

creation of a SQUARE; the attentive knower observes the causes as they make that SQUARE.

How does "putting together" in this way provide causal knowledge? Hobbes's desire to ground geometry in generative definitions provides some indication. A maker first imagines putting four points in motion to make four lines; as Hobbes says: "[. . .] the causes of the properties, that individual figures have, belong to them because we ourselves draw the lines" (OL II.93; Hobbes 1994b, 41). Next they use names (private marks) to form a definition to register their awareness of how they caused the resulting conception. This is why Hobbes's definition for 'line' mentioned already includes a maker's cause in it (MOTION), something Hobbes says Euclid fails to do:

> And where there is place for demonstration, *if the first principles, that is to say, the definitions, contain not the generation of the subject, there can be nothing demonstrated as it ought to be.* And this in the three first definitions of Euclid sufficiently appeareth. For seeing he maketh not nor could make any use of them in his demonstrations they ought not to be numbered among the principles of geometry (EW VII.184; emphasis added).[40]

The definition for 'line' is only part of the entire cause of a SQUARE. Other components are required to make a SQUARE rather than some other figure, and with just one missing, e.g., RIGHT ANGLE, a different figure would be caused, such as a RHOMBUS.

The difference in this use of language and the use of language for prudence cannot be overstated. Rather than private marks arranged into propositions that help a prudent reasoner remember similitudes judged among *many* conceptions (e.g., 'coyote is nocturnal predator'), a definition like 'a line is made by the motion of a point' encodes what a knower observed themselves making in a *single particular* but which applies to *all lines* whatsoever since the definition makes no restriction regarding the quality or quantity of the line. Insofar as the definition records the cause that the knower themselves observed as they brought about the effect, the definition not only helps the knower remember what they brought about ("I made a line by [. . .]") but also helps them remember for future use *how* to make that kind of thing. Definitions like 'line' and the other required components of SQUARE use the most abstract of names like 'motion', which "denotes [*denotare*] the cause of a concrete name, not the thing itself" (Hobbes 1981, 229; OL I.29; EW I.32). The knower observes themselves moving a POINT and that motion is the cause of the concrete name 'moved' and considered without particularities allows the knower

---

[40] See also *De Principiis et Ratiocinatione Geometrarum* (OL IV.421). Jesseph (2021, 61–62; see also Jesseph 1993) discusses Hobbes's criticisms of Euclid on the basis of Hobbes's requirement for generative definitions.

74 HOBBES'S TWO SCIENCES

to register that with the abstract name 'motion'. These names—much more abstract than the names of prudence, such as '*Appietas*' or '*Lentulitas*'—are abstract because they do not include particularities of matter in their account (2012, 58; 1651, 16; cf. Hobbes 1981, 229; OL I.29; EW I.32). I will return to the role of language in section 3.5.

Hobbes's identification of "putting together" as the cause of SQUARE follows from his understanding of causation. He defines 'cause' in *De corpore* IX.3 as an aggregation or sum of all of the accidents *necessary* for some effect: "But a cause simply, or an entire cause [*causa integra*], is the aggregate of all the accidents both of the agents how many soever they be, and of the patient, put together; which when they are all supposed to be present, it cannot be understood but that the effect is produced at the same instant" (OL I.107–108; EW I.121–22). Reassembling a SQUARE by putting these components "together as the cause of the square" (Making Type 1) just is causing a SQUARE; when all are present, a SQUARE *necessarily* is present. A definition of such a figure then helps a maker remember how this was done so it can be done again in the future.

What about the suggested starting point for resolving that I proposed above, i.e., CLOSET DOOR? Although I suggested that the resolution would have begun from such a conception received in sense, which I considered as a SQUARE, I cannot learn the causes of that door *qua* natural body. The causes of many of the door's features, e.g., pattern of woodgrain, are available only to their maker. However, insofar as I consider that CLOSET DOOR as a SQUARE, I can know the reason why it has the shape I take it to have (even if I know it not to be an actually-perfect square) and as such it represents all instances of SQUARE.[41] A maker registers what they discover in making a particular SQUARE with a generative definition, such as the following: a square is made by moving four bodies, considered as points, to equal lengths and assembling those lines together to form four right angles and thus four sides. The maker can remember this entire definition using three names "equilateral, quadrilateral, rectangular," and "for the sake of brevity only" the maker can remember all three names at once with the single name 'square' (Hobbes 1981, 321; OL I.75; EW I.85). This is why "[. . .] definitions in philosophy are prior to the defined names" (Hobbes 1981, 321; OL I.75; EW I.84–85); a knower makes a SQUARE, remembers how they made it with causal definitions along the way, and only then uses names like 'square' to help remember the entire collection of causal definitions. Furthermore, this generative definition is a necessary proposition because, as Hobbes defines 'definition', it is "a proposition, the predicate of which is the resolution of the subject [. . .]" (Hobbes 1981, 319; OL I.74; EW I.83–84), and it will be interchangeable with its hypothetical version (Hobbes 1981, 239; OL I.34; EW I.38–39).

---

[41] Hobbes makes this claim in *Anti-White* IV.2 regarding representation and running water (see fn. 23 above).

SCIENTIFIC KNOWLEDGE FROM MAKING AND THE MECHANICAL MIND 75

Malcolm criticizes Hobbes's claim that actual causes are available to makers but that natural philosophers offer only possible causes as "an inadequate distinction" because, Malcolm contends, it implies that when finding a footprint in the sand— Malcolm's example—one merely "conjectures its cause" but that in "making a footprint oneself" one should also know the actual cause (2002, 155). Hobbes's view would be unsatisfying if it had such a consequence, but there is little reason to think Hobbes would endorse Malcolm's characterization. Even though I am partly the cause of my own footprint, insofar as I *contribute* to its entire cause (OL I.107–108; EW I.121–22), there are many accidents for which I am not the cause that influence the outcome, such as the granularity of the sand, the moisture in the sand, and the mixture of the sand and other components, to name a few. Likewise, I am not entirely causally responsible for a fire I may start but instead am only a single part of a "continual mutation" between agents and patients and can be "considered" as either agent or patient depending upon the explanandum at stake (OL I.109–110; EW I.123–124). In contrast, Hobbes is clear that an artist knows actual causes only when they entirely result from the artist's "own operation" (EW VII.183–184).

A Hobbes-inspired attempt to respond to Malcolm's complaint relies upon the status of the objects used to "make": unlike accelerants for a fire that I desire to make, Hobbesian mathematical objects do not exist *qua* mathematical objects in the world. Hobbes is explicit that there is, for example, no such thing as a line without breadth. The objects of civil philosophy (discussed in chapter 4) have a similar status. When we consider a body as if it had no breadth, Hobbes assumes that we have the complete ability to be the "causes of the properties" that body comes to possess since we are solely responsible for the motion that the line has *qua* artificial body. Thus, although I would be *part* of the total cause of a footprint in the sand or of a body burning upon contact with a flame, perhaps by my moving the flame, there are other parts of that total cause for which I am not responsible (and of which I may have no knowledge apart from suppositional knowledge).

Making thus is a necessary but not sufficient condition for *scientia* according to Hobbes. In some sense, we make footprints, fires, and geometrical objects, but we possess maker's knowledge only for geometrical objects (and, to be discussed in chapter 4, the laws in civil philosophy). The other necessary but not sufficient condition for acquiring maker's knowledge involves the starting conditions for the construction: Hobbes's two instances of *scientia* assume that he is working with a system (*geometric points* or *human bodies*) where he can start the construction without having to attend to any other accidents all. He doesn't argue for this second necessary condition, but it is a consequence of the perspectival principle (discussed more below) and is built into the initial conditions of the two thought experiments he uses (the annihilation of the world and the state of nature thought experiments). We know from experience that we cannot ignore non-geometric features, say, of cakes that we aim to bake, but Hobbes simply assumes that the

76 HOBBES'S TWO SCIENCES

constructions in the sciences of geometry and civil philosophy can begin under conditions where such accidents can be ignored.

Hobbes's account of analysis down to simples and the perspectival principle attempts to make intelligible which features to consider as and which to rule out in instances like footprint making and cake baking. What Hobbes problematically assumes, without argument, is that we simply can "consider as" certain accidents *completely independently* from others (e.g., BODY + MOTION), though not abstracted from matter, and then from such a starting point count as a maker who acquires *scientia*. Locke's distinction in *Essay* III.iii.18 (1975, 418–419) between real and nominal essences would have helped Hobbes be clearer on these aspects of his view, since then Hobbes could have said that in geometry and civil philosophy real and nominal essences coincide, but we do not know this in cases such as closet doors, cake, fire, or footprints (like Locke says of 'gold'). Chapter 5 articulates why Hobbes thought that natural philosophy could still be useful even with this limitation on causal knowledge. Hobbes's natural philosophy held that even though the inability to know actual causes in nature prevents closet door makers or cake bakers from having *scientia* of the outputs of their labor, such individuals can still have suppositional certainty about the causes of those outputs insofar as they borrow causes (the "why") from geometry.

What could license the inferences from *one* particular to *all* of a kind that are implicit along the way as I have made this SQUARE? For example, my definition using private marks to remember "a line is made by the motion of a point" is discovered from moving a single point but applies to all lines universally. A remark Hobbes makes in the triangle example discussed in the next section indicates Hobbes's understanding of this inference: the move from one conception to all conceptions in the case of the triangle is licensed because the conclusion reached was due not to "the length of the sides, nor to any other particular thing in his triangle" but only because the triangle had straight sides and three angles (2012, 54; 1668, 14). Hobbes can make the same claim regarding the definition for 'line': I discover how to cause a line by moving only *one* point, but since the definition makes no reference to the quantity or quality of the line I constructed, it represents all lines *universally*. My definition thus encodes that I *considered as* the LINE that I made in a particular way, i.e., without regard to its length or quality. In encoding the way the mind considered this particular, the definition provides, in Hobbes's words, "an universal notion of the thing defined, representing a certain universal picture [*pictura...universalis*] thereof, not to the eye, but to the mind [*animum*]" (OL I.74; EW I.84).

The same holds for other necessary components of SQUARE such as RIGHT ANGLE and STRAIGHT. Language in the form of a generative definition does not create universality *ex novo*; instead, it registers a discovery made when considering one's conceptual creations as without particularities. Conceptions *qua* conceptions remain particular in the mind, but considering them as without their particularities makes it so their particularity is not registered in a definition. These private marks thus enable us to remember that we learned something universal when considering particulars as without their particularities; like the triangle example, this "makes that which was found true *here*, and *now*, to be true in *all times* and *places*" (2012, 54; 1651, 14). In short, Hobbes can hold that there are no universal ideas/conceptions (EL I.5.6; 2012, 52; 1651, 13; also cf. Hobbes 1981, 277–279; OL I.53–54; EW I.60), as mentioned already, but also claim that *discoveries* of universality are made by "considering as" particular conceptions and then are remembered using language.

3.5 Language for Discovery of *Scientia*

In three major works—*Leviathan* 4 (1651), *De corpore* VI.11 (1655), and Latin *Leviathan* 4 (1668)—Hobbes provides an example of someone who learns that the sum of the angles of a triangle is equal to the sum of two right angles. Hobbes uses this example to distinguish what can be known without using language from what can be remembered with it. Scholarly treatments of this example focus exclusively on Hobbes's invocation of the benefit of language and assume that language alone provides universal knowledge.[42] Indeed, language is crucial to the development of *scientia*—according to *Elements*, names make humans "capable of science" (EL I.V.5). However, scholars have overstated the role language plays for Hobbes in acquiring *scientia*: my view grants that without language the discoveries of *scientia* would be forgotten, but I argue that the mental "observations" of particular conceptions made during constructions are foundational.[43]

---

[42] For example, Peters contrasts Hobbes with Berkeley by observing that the latter used an idea as a sign for all others of a kind but that "Hobbes [. . .] held that universal names are so purely because of their use [. . .]" (1956, 128). Brandt argues that that conceiving universality is only possible by the "aid of names" (1928, 232). Echoing Seth Ward's criticisms, Jesseph locates universality in the imposition of names: "Hobbes had argued that by the imposition of names (such as *triangle* and *equal*) the consideration of a single case could lead to the general result that all triangles have interior angles that sum to two right angles [. . .]" (1999, 218). See also McIntyre (2021, 99). Pettit attributes even greater significance to language, calling it the "magic" that allows humans to move beyond prudence (2008, 25) and arguing that without it there can be no general thought whatsoever and that the advent of language "brings to light [universal] properties that were previously hidden" (2008, 36). Nuchelmans claims that for Hobbes "[u]niversality belongs exclusively to language" and that universal concrete names are private marks of "similar private thoughts" (1983, 137; see also 130–131). But these accounts fail to recognize that for *scientia* language records a single mental observation made of a single conception; it does not to register something among "similar private thoughts." The latter is the function of language for *cognitio* but not *scientia*. de Jong recognizes that discovery is from a single conception but claims that discovery of the triangle rule to be discussed and of other necessary propositions results only from "a language-dependent intensional analysis of a conception of a man" (1990, 80–82). My account, *pace* de Jong, shows that discovery happens only on the basis of contemplating a single conception and that language aids only to "register" what is observed (see fn. 47 of chapter 2).

[43] Hobbes clearly holds that the possession of language is necessary for universalizing in syllogisms: "[. . .] in those animals which lack the use of names no conception or thought answering to a syllogism from universal propositions exists in the mind, since in the course of syllogizing not only is it necessary to think about a thing but also, by alternating changes, about diverse names of a thing, which

78 HOBBES'S TWO SCIENCES

Recall the first "speciall use" of language mentioned earlier: it registers what we discover "by cogitation [1668: *cogitando*]" to be the cause of some effect or what effects some cause may bring about (2012, 50–51; 1651, 13; 1668, 14). I suggest that these references to 'cause' and 'cogitation' refer to the mental act of considering as that involves reaching a foundation and from that foundation constructing to gain causal knowledge. Hobbes's assertion, mentioned already, that this is "in summe, the acquiring of Arts" supports this suggestion and weighs against thinking that this first special use of language concerns prudence (2012, 50; 1651, 13). Furthermore, when we share the causal knowledge that we have remembered with private marks with others using public signs (i.e., the second special use of language, which is the subject of chapter 4), Hobbes uses the very terms we would expect were he talking about *scientia* and not *cognitio*: he says that "we shew to others that knowledge [1668: *ostendamus aliis scientiam*]" by "teaching [1668 Latin: *docere*]" (2012, 50; 1651, 13), the latter of which he elsewhere uses synonymously with demonstrating (e.g., OL I.71; EW I.80–81). This section will apply the account of 'considering as' and construction to the triangle example to show this limited role of language in the acquisition of the universals of *scientia*.

The three instances of the triangle example concern what can be known without the use of language from observing that the sum of the angles of a particular triangle and is equal to the sum of a particular set of right angles. I include these instances in their entirety (with agreements to and changes made in the 1668 Latin edition reflected within the 1651 text).

Leviathan 4 (1651 English and 1668 Latin)

[...] a man that hath no use of Speech at all, (such, as is born and remains perfectly deafe and dumb,) if he set before his eyes a triangle, and by it two right angles, (such as are the corners of a square figure,) he may by meditation compare and find [1668: *meditatione, contemplando comparandoque invenire*], that the three angles of that triangle, are equall to those two right angles that stand by it. But if another triangle be shewn to him different in shape from the former, he cannot know without a new labour, whether the three angles of that also be equall to the same. But he that hath the use of words, when he observes [1668: *animadverterit*], that such equality was consequent, not to the length of the sides, nor to any other particular thing in his triangle; but onely to this, that the sides were straight, and the angles three; and that that was all, for which he named it a Triangle; will boldly conclude Universally, that such equality of angles is in all triangles whatsoever;

---

are applied due to diverse thoughts about a thing" (Hobbes 1981, 261; OL I.45; EW I.50). My claim is that although Hobbes denies that there are any universal conceptions in the mind, the cognitive act of considering as (what I take Hobbes to be referring to in this quotation when he describes thinking about a thing and the "diverse thoughts" one can have about it) allows the human mind to treat a particular conception in a universal manner. Universality, thus, first arises in the mental act of considering as and is secondarily remembered with language.

SCIENTIFIC KNOWLEDGE FROM MAKING AND THE MECHANICAL MIND   79

and register his invention in these generall terms, *Every triangle hath its three an-gles equall to two right angles.* And thus the consequence found [1668: *inventa*] in one particular, comes to be registred and remembred, as a universal rule; and discharges our mental reckoning, of time and place; and delivers us from all la-bour of the mind, saving the first; and makes that which was found true *here*, and *now*, to be true in *all times* and *places*. (2012, 52–54; 1651, 14).

De corpore VI.11

[. . .] if someone, by contemplating [*contemplando*] some triangle put before his eyes, discovered [*inveniret*] that its angles, taken all together, were equal to two right angles, and did so by silently thinking [*tacite cogitando*] the thing without any use of words conceived or expressed, it might happen that he would not know whether or not this property is in another triangle dissimilar to it, or even in the same one viewed in another position. And consequently the contempla-tion [*contemplatio*] would have to be instituted anew for every single triangle presented, and the number is infinite. But this would not be necessary if words were used (of which every universal denotes conceptions of infinite singular things).[44] Nevertheless, they serve for discovering [*inventioni*], as I said before, as tokens for memory, but not as words for signifying. Thus, a solitary man can be a philosopher without a teacher; Adam could have been one. But teaching [*docere*], that is demonstrating [*demonstrare*], requires two people and syllogistic speech. (Hobbes 1981, 311–313; OL I.70–71; EW I.79–80)[45]

---

[44] I have modified Martinich's translation here to side with the Molesworth translation in taking *conceptus* to be plural rather than singular (see also the translation of *Concerning Body*; Hobbes 1656b, 58). Martinich renders "[. . .] (*quorum unumquodque universal singularium rerum conceptus denotat infinitarum*) [. . .]" as "[. . .] (of which every single universal denotes a conception of infinite singular things) [. . .]", but this would require Hobbes to contradict his views, mentioned already, that there are no universal conceptions, i.e., Hobbes could not countenance the view that a word denotes a single con-ception and hold that that conception is of infinite singular things.

[45] In the translator's commentary of *Computatio sive Logica*, Martinich asserts that article 11 is an "intrusion" because "[s]ections 8–10 and 12 deal with the method of uncovering cause" while article 11 "reverts to the doctrine of chapter 2" (Martinich in Hobbes 1981, 431). However, this misses that the triangle example is about discovering the causes (*scientia*) of a triangle by constructing from it and thereby learning its properties. Indeed, the reason the following article, *De corpore* VI.12, concerns "teaching" is that in doing so one is "leading the mind of the one to be taught through the tracks of *our own discovery* [. . .]" (Hobbes 1981, 313; OL I.71; EW I.80). Furthermore, figures in Hobbes's preceding context extensively debated the role and procedures of the "Method of Discovery" (for discussion of figures in Hobbes's intellectual orb, such as Burgersdijk and Keckermann, see Reif 1962, 267ff; 1969). In linking the methods of teaching and discovery together, Hobbes shows affinity to Burgersdijk (see discussion in Hattab 2014; 2021), but Hobbes claims that teaching differs from discovery insofar as it omits the analysis of discovery and that teaching is "demonstration to others" (EW I.80; see Reif 1962, 273, on this point related to Burgersdijk). In actual practice, however, Hobbes uses thought experiments to ground the demonstrative sciences of geometry and civil philosophy (the subject of chapter 4). Given that Hobbes's example of invention is of a geometrical construction, Hobbes differs from Ramus's un-derstanding of invention as a part of logic that sought *arguments* (see Ashworth 1974, 15; Skinner 1996, 59–60). Hobbes's account of "discovery/invention" in construction that I am highlighting also differs from the understanding of "invention" in rhetoric. Skinner pinpoints Hobbes's turn from the *studia humanitatis* and from humanist rhetoric by drawing attention to aspects of Hobbes's translation of Aristotle's *Rhetoric* (Hobbes's *A Briefe of the Art of Rhetorique*) wherein the "translation" undermined

80   HOBBES'S TWO SCIENCES

These accounts of the triangle example in the English (1651) and Latin *Leviathan* (1668) and in *De corpore* (1655) present essentially the same view. Hobbes claims in each rendition that a human without the use of language could examine a triangle placed in front of them and successfully determine the relationship of the sum of its angles to the sum of the two right angles beside it. How would such a comparison work? In *De corpore* VI.11, Hobbes describes the mental activity as a "contemplation [*contemplatio*]" that happens by "silent thinking [*tacite cogitando*]." He uses similar terms in *Leviathan* 4, calling it something that happens "by meditation" whereby without language one can "compare and find [1668: *comparandoque invenire*]" the equality.[46] However, to the 1668 Latin Hobbes provides a connection to the text of *De corpore* by adding *contemplando*; thus, the mental activity described in 1668 is as follows: "[. . .] by meditation, contemplating and comparing find [. . .]."[47] Furthermore, with the addition of "eventually [*tandem*]" in 1668 Hobbes suggests that this effort may be difficult without language (2012, 53; 1668, 15) but nevertheless possible.

Hobbes describes the result of this non-linguistic "contemplation" as a "discovery" in all three versions, and he uses multiples instances of forms of *invenire* in *De corpore* and the Latin edition of *Leviathan*. Indeed, the example in *De corpore* VI.11 is embedded within a discussion of the "method of discovery" (Hobbes 1981, 311; OL I.70; EW I.79; see fn. 45 above). Finally, both accounts emphasize that without private marks "what we discover [*invenimus*] perishes" (OL I.70; Hobbes 1981, 311) and there will have to be "a new labour" (2012, 52; 1651, 14). Hobbes's brief remarks much earlier in the Fourteenth Objection to the *Meditations* support this account of non-linguistic, conceptual discovery. There Hobbes differentiates when "we have conceived in our thought that all the angles of a triangle add up to two right angles" from the subsequent activity of naming wherein "we bestow on

---

Aristotle's own claims and showed Hobbes's suspicions of rhetoric and eloquence (Skinner 1996, 256–257), e.g., expressing views of rhetoric that sound more like the Platonic view of rhetoric as cookery than Aristotle's own view. This critique, Skinner argues, continued in *Elements* and *De cive*, and is reflected in Hobbes's focused attacks on *inventio* in rhetoric (see Skinner 1996, 257–267). Raylor likewise focuses on Hobbes's criticisms of *inventio*, but locates the indictment of *inventio* not as a turn from rhetoric but rather of its use in philosophy, since the (rhetorical) method of invention allows the use of mere *endoxa*, perhaps undefined or lacking clarity, to serve as a foundation for theorizing in political philosophy (Raylor 2018, 223–224).

[46] Hobbes uses 'meditation' as a synonym for 'contemplation'. For example, in *De corpore* XXVII.1 when describing how "after meditation and contemplation [*post meditatione et contemplationem*]" things that seemed previously unimaginable (such as the immense distance between the fixed stars and Earth) become familiar and "we believe them," and he suggests that likewise immensely small bodies will eventually be believed to exist by many (OL I.364; EW I.447). See fn. 49 below for more details regarding Hobbes's other uses of 'contemplation'.

[47] The addition of *contemplando* is represented on the Latin page of the 2012 edition of *Leviathan*, but unlike the other additions to this passage, *contemplando* is not reflected in the apparatus (e.g., *tandem* is represented as an addition; cf. 2012, 52 fn. 13).

SCIENTIFIC KNOWLEDGE FROM MAKING AND THE MECHANICAL MIND   81

the triangle this second label 'having its angles equal to two right angles'" (CSM II.135; AT VII.193).

What happens in common for the two laborers, one without language and the other with it? It may seem Hobbes thinks that the "contemplation" by the non-linguistic individual would be merely sense based—the individual silently thinking without language compares the imagistic conception of the TRIANGLE and the imagistic conception of TWO + RIGHT + ANGLE received in sense and stumbles upon their equality. However, this understanding of Hobbes's reference to "contemplation" (in the Latin versions) and "meditation" (in the English *Leviathan*) would make the non-linguistic individual's inference analogous to the judging of similitudes from conceptions received in sense for prudence (*cognitio*). There are three reasons to resist such an understanding of non-linguistic "contemplation." First, Hobbes describes this observation as resulting from "labour" and emphasizes that even the non-linguistic individual *discovers* something. Second, elsewhere in *De corpore* I.5 Hobbes claims that if one were to examine a circle one could not know by "sense" (*sensu*) whether it is a circle, i.e., by sense one could not know the properties of the figure at all (Hobbes 1981, 181; OL I.5; EW I.6). In contrast to such sense-based examination of a figure, Hobbes argues that determining its properties "from the known generation of the displayed figure, [is] most easy" (Hobbes 1981, 181; OL I.5; EW I.6).[48] Third, since the observation of equality requires a comparison of the sum of three discrete angles to two discrete right angles, imagistic conceptions could not be compared without the construction of *another* figure—a circle—to measure and sum up the angles so that comparison is possible.

Given these reasons, I suggest that both the individual with language and the individual without language would engage in contemplation/meditation by constructing from the triangle placed before them by drawing additional lines.[49]

---

[48] Thus, Hobbes would reject Seth Ward's claim that the equality between the particular triangle and two particular right angles could be discovered either "by experience [*experientia*] (so much as the equality of things can be known by sense [*sensu*]); for example by aid of a pair of compasses measuring the angles of the arcs of circles; or it can be found by a deep and attentive cogitation [*cogitatione*] of the soul" (Ward 1651, 31; trans. in Jesseph 1999, 218). Although Hobbes holds that one should use a drawn circle and compass to measure the quantity of angles (discussed below), since this involves *constructing* a circle (as well as constructing lines from the initial triangle) this would not count, in Hobbes's view, as knowing by sense but rather as knowing by making (*scientia*). In *Six Lessons*, Hobbes claims that if someone had never seen a circle made then it would be difficult to convince him that "such a figure [is] possible" (EW VII.205).

[49] Hobbes uses the terms 'contemplation' and 'meditation', the latter of which occurs in the English *Leviathan*, as terms of art referring to the cognitive activity of philosophizing generally and of the sciences of geometry and civil philosophy in particular. Hobbes also uses 'physical contemplation' when referring to the cognitive activity of natural philosophy. Regarding philosophizing generally, he asserts that the order of human "contemplating" (*contemplandi ergo ordo erit* [. . .]) should follow the order of God's creation (*creandi ordo erat* [. . .]) in Epistle to the Reader of *De corpore* (1655, A4v; OL I.unnumbered page). In the Epistle Dedicatory of *De cive*, prior to identifying philosophy with geometry, physics, and "morals," Hobbes declares that philosophy begins with "contemplation" of particulars: "For Philosophy opens the way from the contemplation [*contemplatione*] of individual things to universal precepts" (OL II.136). Tuck and Silverthorne (Hobbes 1998, 4) obscure

82 HOBBES'S TWO SCIENCES

Then they would compare their construction with the two right angles by using a constructed circle. Hobbes provides no details about what such "labour" by "contemplation" would amount to in the examples above, but he does provide an example of just such a construction in *De corpore* XIV.12, Corollary IV, where he demonstrates that "the three angles of a straight-lined plain triangle are equal to

the cognitive act at the beginning of philosophy by translating *contemplatione* as 'observation' (cf. OL II.141–142; Hobbes 1998, 7).

In addition to the triangle examples discussed above, Hobbes elsewhere uses forms of 'contemplation' to refer to the form of cognition characteristic of geometry. For example, he describes geometry as arising from "contemplation [*contemplatione*]" wherein one constructs through a compositive method (*methodus compositiva*) from a single body, to "simple motion" by that body making a line, to motion of a line to generate to a surface and so on (see *De corpore* VI.6; OL I.63; EW I.71). Likewise, he uses 'contemplation' to refer to the parts of his own geometry that consider motion generated by another body's motion (OL I.63; EW I.71–72), which includes chapters in *De corpore* Part III such as chapter XXI on circular motion and chapter XXIV on reflection and refraction. See also his assertion in Lesson I of *Six Lessons* that the "science of geometry [...] contemplateth Bodies onely [...]" (1656a, 2; EW VII.193) and his mocking tone in Lesson IIII (*sic*) that his opponent has erred because he "cannot yet, nor perhaps ever will contemplate Time and Motion (which requireth a steddy brain) without confusion" (1656a, 35; EW VII.280). See also uses in *Examinatio et Emendatio* (OL IV.84, 96, 222).

Regarding civil philosophy, Hobbes criticizes Aristotle's "axiom" that humans are political animals as false and states that it "proceeds from a superficial contemplation of human nature [*a nimis levi naturae humanae contemplatione*]" (OL II.159; Hobbes 1998, 22). Potentially weighing against seeing 'contemplation' as a Hobbesian term of art referring to cognitive activity behind philosophizing, especially for *scientia*, are two uses in the English version of *Leviathan* that refer to sense-based *cognitio*: in *Leviathan* 7 when describing beginning a discourse from "some other contemplation" and not a definition results in "opinion" (2012, 100; 1651, 31) and in *Leviathan* 13 to refer to those who in the state of nature take "pleasure in contemplating their own power in the acts of conquest" (2012, 190; 1651, 61). However, these are minor deviations and these occurrences of 'contemplating' in the English version are not reflected in the Latin version (cf. 2012, 101, 191; 1668, 33, 64). Furthermore, in the English *Leviathan* Hobbes frequently uses the synonym 'meditation' (see fn. 46 above) to refer to the same cognitive activity as 'contemplation'. For example, he warns against "men that take their instruction from the authority of books, and not from their own meditation" (2012, 56; 1651, 15). He claims in *Leviathan* 30 that "deep meditation" is required for "learning of truth, not onely in the matter of Naturall Justice, but also of all other Sciences" (2012, 532; 1651, 179). Hobbes furthermore describes a good counselor as one who in addition to having experience has "much meditated on" the science of civil philosophy and possesses "great knowledge of the disposition of Man-kind, of the Rights of Government, and of the nature of Equity, Law, Justice, and Honour, not to be attained without study" (2012, 406; 1651, 134). Finally, *Leviathan* 26 describes "a good Judge" as one who has had the most leisure and the greatest inclination to meditate, adding in 1668 that this inclination is "to the science of the equitable and good [*ad scientiam Aequi & Boni*]" (2012, 438; 1651, 146–147; cf. 2012, 439 fn. 79; 1668, 135), and in *Leviathan* 30 he differentiates failed commonwealths from commonwealths that are "everlasting," unless failing from "externall violence," as being rooted in the latter being founded upon "Principles of Reason" that are "found out, by industrious meditation" (2012, 522; 1651, 176).

Hobbes similarly uses contemplation/meditation in describing natural philosophy. When stopping a potential regress into additional mental faculties in *De corpore* XXV.1, Hobbes uses the cognate verb *contemplare* in the mouth of his imagined interlocutor who asks by what sense would sense *take notice* of sense, to which Hobbes replies by sense itself, of course (OL I.389; EW I.317). He later uses 'contemplation' to describe the activity of natural philosophy in contemplating sense itself first and then bodies generally (see OL I.334; EW I.410). Hobbes furthermore differentiates natural-philosophical from geometrical contemplation by calling it "physical contemplation [*contemplationis physicae*]" in two places: first, when explaining that "moral philosophy" must be considered after physics because the former has its causes in sense and imagination, which "are the subject of physical contemplation" (OL I.64; EW I.73); and second, at the end of *De corpore* in chapter XXX he emphasizes that Part IV has offered possible causes not actual causes, "which is the end of physical contemplation" (OL I.431; EW I.531). He similarly uses 'contemplation' to describe natural-philosophical activity in *Elements* (EL I.IX.18), identifying it as driven by curiosity that varies by degrees among knowers. See also Hobbes's

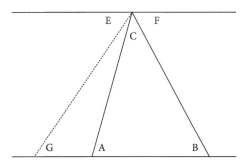

Figure 3.1 Representation of a Geometrical Construction in *De corpore* XIV.

two right angles" (OL I.165; EW I.109). To my knowledge, no other discussions of the triangle/right-angles examples in the literature have made this connection to the demonstration in *De corpore* XIV.12.

Hobbes uses the capital letters on parts of the triangle above (see Figure 3.1) as names within the proof (see *De corpore* II.4 on single letters as names in geometry; Hobbes 1981, 199; OL I.14; EW I.16), and the demonstration Hobbes provides follows from his "universal" definition of parallel lines: as "any two lines whatsoever, strait or crooked, as also any two superficies, are parallel; when two equal straight lines, wheresoever they fall upon them, make always equal angles with each of them" (OL I.163; EW I.189).[50] The demonstration relies upon two additional claims: first, the quantity or measure of an angle is determined by "its proportion to the whole perimeter" of a circle (OL I.161; EW I.186); and second, that a right angle is defined as an angle measuring one fourth of the circumference of a circle (OL I.162; EW I.187).

I will first discuss what the individual *without language* does to discover the equality of the sum of the angles of the particular triangle to the sum of the particular right angles, and then I will discuss how the individual with language uses private marks to register, in the end, that "*Every triangle hath its three angles equall to two right angles.*" As mentioned already, Hobbes thinks sense alone would provide no help in comparing the triangle with the two right angles, so the individual without language might try many things, possessing leisure and driven by curiosity to see all of the effects that are possible by manipulating this geometric figure by

---

reference to the "Contemplation of Nature [*Naturae . . . contemplatio*]" in "To the King" of *Seven Philosophical Problems* (1662, sig. A3v; 1682, sig. Av). Hobbes asserts that astrology tries to mimic natural philosophy by "contemplation [*contemplatione*]" of the stars and from that predict the future, but he denies it the status of *scientia* (*non scientiae est*) (OL II.127). See also *Decameron Physiologicum* (1678, 2, 14).

[50] For discussion of Hobbes's definition of 'parallel' and his responses to criticisms from Claude Mylon, see Jesseph (1999, 253–254).

84   HOBBES'S TWO SCIENCES

adding to it or taking away from it. Some of these leisurely endeavors would presumably amount to nothing. However, to be successful in the end that interests Hobbes here, they would eventually need to extend—by constructing—one of the sides of the triangle with a line that is continuous to it and extends beyond its angles. Likewise, the individual would draw a line parallel to the line extended from a sides. Finally, the individual would trace the original lines of the triangle on all sides, thus having in the mind the conception—not the name—of THREE and then considering the sides as STRAIGHT + LINE.

Next the individual without language who is successful at this attempt would draw or imagine a circle and compare the proportion of the circumference of that circle they have drawn taken up by one angle of the triangle and two of the alternate angles created by the parallel lines with the proportion taken up by the two right angles considered together. We can only make sense of such activity for the curious person who has sufficient leisure. Upon observing that the two right angles take up the same proportion of the circle as the single angle of the triangle and two of the alternate angles, the individual without language has discovered that the angles of *this* triangle are equal *these* two right angles.

What does language add to this picture? In the *Leviathan* 4 rendition, Hobbes makes the distinction clear: "But he that hath the use of words, when he observes [1668: *animadverterit*], that such equality was consequent, not to the length of the sides, nor to any other particular thing in his triangle; but onely to this, that the sides were straight, and the angles three; and that that was all, for which he named it a Triangle." I suggest that this ability to observe the reason for the equality is present for the individual without language just as well as for the individual with language. Hobbes's use of *animadvertere* in the 1668 *Leviathan* version of the example supports this understanding, since elsewhere Hobbes uses this term and cognates to refer to the cognitive activity of *noticing* features about one's conceptions without any mention whatsoever of language.[51] One notices similarities and differences among a group of conceptions received in experience when judging similitudes to forming the universals of prudence (*cognitio*). These two individuals described in the triangle example, however, have examined only a single triangle, a pair of right angles, and a single constructed circle. From that they both can determine the sums of the respective angles to be equal by comparing the proportion (not particular measurement) of the circumference of the circle that they occupied. They are both capable of observing that the equality rested *only* in the fact that the triangle that they had traced and extended lines from had three sides and straight sides: they cannot gain this from sense, since, as mentioned already, Hobbes denies

---

[51] For example, in *De corpore* XXV.8 Hobbes declares that "[...] he that thinketh [*cogitat*], compareth the phantasms that pass, that is, taketh notice [*animadvertit*] of their likeness or unlikeness to one another [...]" (OL I. 325; EW I.399). Likewise, in *De corpore* VI.2 Hobbes describes how the whole idea of a man is known before "we notice [*advertimus*]" the particular ideas, such as the idea of ANIMATE (Hobbes 1981, 289; OL I.59; EW I.67).

that the properties of figures can be known by sense (*sensu*). In other words, the individual without language is just as capable as the other in learning the reason for the equality in proportion of the circle's circumference—the problem according to Hobbes is that, due to the amount of effort needed to reach that point, the individual without language will forget the details of their discovery and have to repeat the labor for each triangle they encounter.

Language users are thus able to register that they *considered* the particular triangle *as* without respect to the particular lengths of its sides. The mental act of considering as provides a foundation for construction (drawing additional lines and a circle) that removes the particularity of the triangle from consideration and then the attentive knower with language can remember the conclusion of their labor with a necessary proposition (cf. EW I.40) that registers the property discovered: "*Every triangle hath its three angles equall to two right angles.*" The steps would proceed as follows:

The individual *with language* would ...

1. trace the triangle presented and, considering it as a triangle without respect to the particular lengths of its sides, register this with a necessary proposition in the form of a definition: A triangle is made by moving three points to form straight lines, and joining them together to form three angles;
2. label the sides considered as straight lines with letters acting as names, such as A, B, and C, the act of which makes it easier to ignore their individual lengths;
3. draw a line by extending A–B and then draw a new line E–F;
4. draw a circle, H, without its particular diameter or circumference being considered;
5. measure the quantity of the sum of $\angle ECA$, $\angle ACB$, and $\angle BCF$ and compare the proportion (not measure) of the circumference of the constructed circle H taken up by E, C, and F with the proportion of the circumference taken up by two drawn right angles;
6. "observe" that the equality in terms of proportion depends only upon the particular triangle being considered *as* a triangle and nothing else; and
7. declare that the equivalence—a property discovered—holds for all triangles considered as such with the universal rule, which is a necessary proposition.[52]

---

[52] Criticizing appeals to formal causation (e.g., the figure of the triangle is the *formal* cause of the property that its angles are equal to two right angles), Hobbes claims that the properties of a geometrical figure, like the property that "a figure has all its angles equal to two right angles," do not follow from that figure but rather are present "simultaneously with it" (Hobbes 1981, 247–249; OL I.38–39; EW I.43–44). He admits that one may know the figure before knowing the property but denies that there is any other sense in which the one precedes the other. Malcolm notes that for Hobbes a moving point counts as the "cause" of a figure but states that Hobbes denies that the *way* in which those motions occur does not count as such, asserting that Hobbes would deny that "the equidistance of the resulting line from the central point" is part of the cause of a circle (Malcolm 2002, 155). Hobbes does clearly deny that formal causes are independently responsible for the properties of figures, but this need not prevent him, *pace*

86 HOBBES'S TWO SCIENCES

Just like Hobbes states that the surface of running water "represents [*repraesentare*] not that particular water but any water, or air, or body of the same size and shape" (1973, 126–127; 1976, 53), according to Hobbes, what is learned about triangle ABC represents all triangles *qua* triangles.[53] Importantly, language provides neither the ability to *construct* (and gain *scientia*) nor the ability to "observe" the equality in the triangle example. Instead, private marks provide a discoverer only with a tool with which to remember discoveries, which helps facilitate further discoveries. Even lone Adam could have done this and insomuch become a "philosopher without a teacher" (Hobbes 1981, 313; OL I.71; EW I.80). In short, language enables us to register what we discover "by cogitation [1668: *cogitando*]" to be the cause of some effect or what effects some cause may bring about (2012, 50–51; 1651, 13; 1668, 14). The private mark 'triangle' thus "denotes conceptions of infinite singular things" (Hobbes 1981, 311–313; OL I.70–71; EW I.80), the set of all triangles having been constructed or that could be constructed.[54] Whenever lone Adam uses the mark 'triangle', a particular TRIANGLE will come to his mind, one of a determinate size and having a particular arrangement of angles that sum to two right angles, but insofar as Adam has registered his observation correctly, that mark 'triangle' also helps recall the particular way in which Adam considered that triangle as without any regard to its particular size or arrangement of angles (were those features registered, the rule would instead begin "every triangle with sides measuring [. . .]"). In other words, the mark registers what Adam "observed" when he *considered as.*

Since my focus is the acquisition of *scientia* and the role language plays in retaining it, which has been neglected by Hobbes scholarship, I leave to the side debates concerning Hobbes's broader philosophy of language. Abizadeh (2015) taxonomizes pre- and post-1960s interpretations of Hobbes's philosophy of language. Nuchelmans argues that Hobbes breaks with the "traditional frame of thought" that distinguished between referring to things (denotation) and ways of conceiving them (connotation) and that Hobbes "shifts the universality which is involved in referring expressions from the conception to the name." Nuchelmans further argues that for Hobbes "the concrete name which is applicable to more than one body [. . .] is the bearer of universality," not a conception (Nuchelmans 1983,

---

Malcolm, from including *ways that motion happens* as parts of the cause (as he clearly does in the case of SQUARE discussed above).

[53] Duncan argues that although Hobbes denies that there are universal ideas in the mind or universals in nature, he fails to consider that ideas might be universal "because they themselves represent multiple things" (2017, 51; also cf. Laird 1968, 147ff). My account shows, *pace* Duncan, that this is exactly what Hobbes endorses, not in the nature of ideas themselves but rather in the mental *act* of considering as.

[54] Just like the examples of positive names 'man' and 'philosopher' Hobbes provides in *De corpore* II.7, the name 'triangle' denotes any one of many triangles (Hobbes 1981, 203; OL I.16; EW I.18). When used within a definition or a universal rule about one of its properties, 'triangle' directs one to consider as whichever particular of these many that is brought to mind without its particularities.

SCIENTIFIC KNOWLEDGE FROM MAKING AND THE MECHANICAL MIND   87

130–131; see also de Jong 1986, 126–127). However, the triangle example properly understood shows, *pace* Nuchelmans, that for *scientia* universality is *discovered* by considering a particular conception as without any of its particularity and that the name 'triangle', whether as a mark or sign, simply aids in remembering that cognitive act, i.e., the observation that Hobbes calls a "discovery." Nuchelmans's view that "the set of individual conceptions or ideas of which a name is a sign" is "[w]hat lends meaning to a universal concrete name" certainly holds for cases of *cognitio*, for which there will be sets of conceptions received in experience which can be considered in different ways, but it does not apply for Hobbesian *scientia*, which is grounded in a *single* conception whose properties are known because of a particular construction but which applies universally to all of that kind.

My account thus shows how Hobbes would answer J.S. Mill's worry in *System of Logic* IV.2 (mentioned already in chapter 2) that Hobbes's account of propositions applies only to proper names and "is a sadly inadequate theory of any [other names]" (1974, 91). Mill's diagnosis is that

> Hobbes, in common with the other Nominalists, bestowed little or no attention upon the connotation of words; and sought for their meaning exclusively in what they denote: as if all names had been (what none but proper names really are) marks put upon individuals [...]" (1974, 91).

de Jong raises related worries about the relationship between universals and names of particular things and Hobbes's view that a name signifies a single conception (1990, 77–78). Duncan argues that Hobbes's account of universals denies "conceptualism as well as realism" and holds that there is "no special sort of idea that is associated with general names but not with proper ones" (2017, 51).

The account I offer shows Hobbes's answer to these various worries: the name 'triangle' denotes all the triangles, but as a mark used to remember one's discovery it registers that one considered one of the triangles being denoted in a particular way (considered as only *triangular*). When using 'triangle' in the future, any one of the denoted triangles may be called to mind, but if the mark functions properly it will not matter which one is recalled since it will be considered as just triangular. In Mill's terms, though not the phrasing used by Hobbes, the name 'triangle' in the rule remembered connotes the accidents for which the name applies, i.e., simply having three sides and three angles (see Mill's example of the connotation of 'man'; *System of Logic* IV.2; 1974, 92).

3.6 Hobbesian Perspectival Epistemology

This chapter began by describing two criteria of Hobbesian *scientia*: first, *scientia* provides the bearer with causal knowledge; and second, *scientia* is for the sake of

88   HOBBES'S TWO SCIENCES

power (*potentia*). The focus of the chapter so far has been on the first criterion; on the basis of the causal knowledge requirement, Hobbes distinguishes *scientia* from prudence and all other forms of *cognitio*. Recall also Hobbes's declaration in *De corpore* I.6 that *scientia* provides benefits to its holders: "[k]nowledge is for the sake of power [*Scientia propter potentiam*]" (Hobbes 1981, 183; OL I.6; EW I.7). Hobbes describes these benefits as the "comforts" stemming from knowledge of geometry and its uses in physics, such as marking "moments in time" and "mapping the face of the earth" (Hobbes 1981, 183–185; OL I.6–7; EW I.7). Civil and moral philosophy provide benefit because of what they *prevent*, as Hobbes argues that "civil wars and then the greatest calamities follow upon ignorance of political responsibilities, that is, moral science [*moralis scientiae*]" (Hobbes 1981, 187; OL I.8; EW I.9–10).

Thus, in terms of the "power" criterion, geometry and civil philosophy count as *scientiae* because Hobbes assumes that all humans desire to avoid war, that is, to have peace, to have the things necessary for commodious living, and the hope that they can achieve those things (2012, 196; 1651, 63). The first desire is responsible for the advent of civil philosophy, and the second and third are responsible for geometry and its use in natural philosophy. Further requirements are the drive of curiosity and presence of leisure. The chapter so far has shown how Hobbes thought causal knowledge was possible in the science of geometry. How the maker's knowledge principle is satisfied in civil philosophy will be the subject of chapter 4.

However, I suggest that ends desired by all humans not only make these two domains of inquiry *scientiae*, but they also inform the work done *within* those sciences. This occurs by the influence that knowers' interests and aims have upon how they use the cognitive device I have referred to as 'considering as'. Both of the ways that perspective plays a role in Hobbes's epistemology can be explained by what I call the perspectival principle:

> **The Perspectival Principle**: *explananda* are determined by human interests and aims, and an *explanadum* determines which *explanans* is relevant

This chapter concludes by briefly considering how perspective plays this role by drawing attention to the stopping point of the analysis to humans' natural state and to the flexibility of the use of geometry within Hobbesian natural philosophy.

3.6.1 Perspective Stops an Analysis

The analysis of SQUARE terminated in what Hobbes identifies as "our simplest conceptions" in *De corpore* VI.6. For the purposes of that example, the simplest conceptions mentioned were BODY, PLACE, and MOTION. In addition to these, Hobbes includes a distinction between imaginary SPACE and real SPACE: the former encodes exteriority in any conception caused by an external body (for discussion, see Slowik 2021), and the latter is co-extensive with PLACE. Hobbes regards these conceptions as simple because they have no further parts than can be considered and, furthermore, because they are the necessary components for *all* conceptions. It is impossible to imagine any conception that does not contain these conceptions.

The foundation of the science of geometry is thus established by the nature of conceptions themselves. There can be no further resolution of conceptions since any attempt to subtract BODY or PLACE would result in no conception at all. Even if one *desired* to resolve further, perhaps to a conception putatively signified by the mark 'incorporeal substance', there could be no doing so in a way that preserved a conception. Remove BODY and any imagistic conception disappears. This is why Hobbes contends that phrases like 'hypostatical' are absurd (2012, 72; 1651, 20–21). Hobbes declares in *Elements* that one who uses such phrases lacks "evidence," i.e., the words used have no concomitance with conceptions (EL I.VI.3).

The analysis, or resolution, of conceptions for geometry is thus stopped by itself since one cannot remove conceptions like BODY from consideration and have anything remaining in the mind. However, unlike geometry, the foundation of the science of civil philosophy is not determined by the nature of the conceptions under consideration. Why stop the analysis with humans in their natural state? Why not resolve further into BODY, considered generally, and MOTION (the subjects of natural philosophy)? Some scholars have argued that Hobbes held that the analysis *could* continue beyond humans in their natural state and have sought to understand Hobbesian civil philosophy as part of a reductive physicalism in which claims about morality ultimately reduce to claims about microscopic bodies in motion (e.g., Grene 1969; Hampton 1986). There are many reasons to reject such a view (e.g., Adams 2016a; 2021; Malcolm 2002, 147), as discussed in chapter 1, but the account of Hobbesian epistemology I have sketched shows why the reductive physicalist account cannot be Hobbes's considered view. Indeed, were it correct, there could be no *scientia* of civil philosophy since, as mentioned already, according to Hobbes the causes of natural phenomena are not accessible to humans. Thus, civil philosophy would not satisfy the maker's knowledge principle and would fail to count as *scientia*.

Furthermore, Hobbes's civil philosophy never offers such a reduction. According to the maker's knowledge understanding of Hobbesian science, Hobbes's major works—*Elements*, *De cive*, and *Leviathan*—begin the science of civil philosophy by considering humans in their natural state. While *Elements* and *Leviathan* treat human psychology in the chapters that precede the discussion of humans' natural state, the content from those chapters is not directly imported into the construction that begins from the state of nature and ends with the making of PEACE using the Laws of Nature. For example, Hobbes's detailed natural-philosophical account of sensation, imagination, memory, and experience plays

90   HOBBES'S TWO SCIENCES

no significant role in the making of PEACE (discussed in chapter 4). This division is clear when, in *De corpore*, the explanation of sense is in Part IV on natural philosophy (discussed in chapter 5). *Elements* I.14.1–2 implies this difference between Hobbes's natural philosophy and civil philosophy projects: on the one hand, in the natural-philosophical account of human psychology in the preceding chapters, Hobbes says he has offered "the whole nature of man," while on the other hand, in the chapter on the natural state, he notes that it is "expedient" to begin by considering humans less comprehensively in "mere nature."

What could license such a move rather than a full reduction to natural bodies, whether through analysis or a change in the starting point of the thought experiment? Why make this the stopping point where one begins civil philosophy? Unlike in geometry, where the nature of conceptions themselves provides the stopping point of the analysis, I suggest that in the case of civil philosophy the aims and interests of knowers fix the stopping point. The usefulness of civil philosophy, as Hobbes notes in *De corpore* I.7 (Hobbes 1981, 187; OL I.8; EW I.8), is its ability to prevent war. Given that the cessation of war (i.e., peace) is desired by all humans, in initiating an analysis Hobbes can begin by considering a conglomerate particular conception HUMAN + WAR and abstract away WAR by seeing contained in it, among others, the particular conception CIVIL RELATIONS. Further analysis of the conception HUMAN reveals it contains BODY + ANIMATE + RATIONAL. Examining the conception ANIMATE reveals that it contains VITAL MOTION and VOLUNTARY MOTION. None of these conceptions contained in HUMAN can be removed from consideration if the aim is to understand *how* HUMAN was joined together with WAR in the first place, the latter of which includes CIVIL RELATIONS, but WAR can be removed; indeed, the very goal of civil philosophy is to learn precisely *how* to prevent WAR from occurring. Since we remove WAR, and thus CIVIL RELATIONS, HUMAN in its natural state remains. This stopping point for an analysis (or, in the case of Hobbes's actual presentation, the starting point for a synthetic demonstration) is thus determined by the goal. Given the perspectival principle, since peace is the *explanandum*, this determines that the *explanans* should be at a level that will show how to bring about peace.

Hobbes's descriptions of the starting point of the science of civil philosophy fit with this account. After praising geometers for having "managed their province outstandingly," in the Epistle Dedicatory of *De cive* Hobbes faults civil philosophy and claims that "there is no greater knowledge [*scientia*] of natural right and natural laws today than in the past" (Hobbes 1998, 4–5; OL II.137–138). The reason for such failing, Hobbes claims, is a lack of a proper starting point:

> [t]he single reason for this situation seems to be that none of those who have dealt with this subject have employed a suitable starting point from which to teach it. For the starting point of a science [*scientia*] cannot be set at any point we choose as in a circle. (Hobbes 1998, 5; OL II.138)

The solution to this predicament is to find the correct place to begin the science of civil philosophy. This leads Hobbes in *De cive* not to appeal to a microscopic level but instead to what he identifies as "two absolutely certain postulates of human nature": greed understood as desire to have private use of common property, and natural reason understood as striving to avoid death (1998, 6; OL II.139). From these starting points, which are properties of HUMAN and not BODY in general, Hobbes claims to have demonstrated the "elements of moral virtue and civil duties."

Likewise, he concludes the first part of *Elements* (Humane Nature) immediately following the final article of chapter XIII with a summative claim: "Thus we have considered the nature of Man so far as was requisite for the finding out *the first and most simple elements wherein the compositions of Politick Rules and Laws are lastly resolved*; which was my present purpose" (1650, 170; emphasis added).[55] Finally, the state of nature thought experiment in *Leviathan* 13 begins by considering humans devoid of CIVIL RELATIONS and with features of HUMAN remaining, not BODY considered generally.

Hobbes's non-reductive approach is intuitively plausible. If one's goal were to explain human digestion or some lower-level phenomenon, then one would appeal to some "lower" level for the simples. Indeed, that is precisely what Hobbes does in the optics of *De homine* II when explaining visual perception by appeal to the parts of the eye and to geometrical principles like those related to reflection and refraction which require considering parts of the eye as if they were geometrical objects (Adams 2014b). Likewise, if one wanted to explain why a human body falls when slipping from the edge of a cliff, one would consider that entity not *qua* HUMAN BODY but *qua* BODY, and other simples would ground the explanation. Are these examples, and those from civil philosophy above, decisive in refuting a reductive physicalist interpretation of Hobbes? No, they are not decisive, and Hobbes himself seems to grant that civil philosophy is deriv*able* from first philosophy and natural philosophy while also holding that "it can nevertheless be detached from it" (Hobbes 1981, 301; OL I.65; EW I.73). Clearly, given Hobbes's understanding of analysis, the conception HUMAN BODY considered in the natural state *could* be resolved into simpler conceptions, but the perspectival principle, not the reductive physicalist view, provides a way to understand the *actual* structure of Hobbes's major works of civil philosophy.

Even if they are not decisive, understanding the foundation of the *scientia* of civil philosophy in this way places the burden of proof upon the reductive physicalist view. Not only would the reductive physicalist account fail to make sense of the causal criterion for *scientia*, since Hobbes is clear that humans cannot know the causes of *natural* bodies, whether human or otherwise, but it would also fail to make sense of the criterion that *scientia* is for power and thus must have utility.

---

[55] Tönnies provides this "conclusion" in a footnote (Hobbes 1928, 191 fn. 11).

92   HOBBES'S TWO SCIENCES

Say that we continued the analysis of HUMAN beyond BODY + ANIMATE + RATIONAL, and from there resolved ANIMATE to MOTION (i.e., MOTION or REST) and simply abstracted RATIONAL away to get to some bedrock, as it were, of reality. What use would such a resolution have for those who want to avoid war? It would have no use whatever, since in synthesis one would have to add back in the conceptions that are *uniquely* human. In other words, Hobbes thinks the analysis can terminate at HUMAN since that provides a sufficient starting point to gain what humans desire—PEACE. This account will be given more attention in chapter 4.

3.6.2 Perspective Determines Use of Geometry within Natural Philosophy

The perspectival principle likewise explains Hobbes's uses of geometry in natural philosophy. Hobbes recognizes from earlier works like *De cive* that natural philosophy should use geometrical principles in its explanations. He claims that "[. . .] whatever in short distinguishes the modern world from the barbarity of the past, is almost wholly the gift of Geometry; for what we owe to Physics, Physics owes to Geometry" (Hobbes 1998, 4–5; OL II.137). In the natural philosophy section of *De corpore* (Part IV) and in other works such as *Dialogus Physicus, sive De natura Aeris*, Hobbes provides citations to prior chapters in which the demonstration of this or that principle occurred (discussed in chapter 5).

Hobbes's understanding of the mental act of considering as allows a great deal of flexibility for the ways in which conceptions can be treated. This ability to "consider as" allows a human knower to take any conglomerate conception received in sense, or derived from those that have been, and subtract from it, ignoring certain features and focusing exclusively on others. For example, as has been mentioned already, concerning geometrical objects Hobbes denies that breadthless lines exist but holds instead that there are bodies that we "consider as" having no breadth (EW VII.202). When *using* geometry within natural philosophy, Hobbes similarly holds that a knower may consider a conception of an external body in different ways, depending upon their aims and interests. For example, if a natural philosopher is interested only in its annual path, they may "consider" the Earth as if it were a single point (OL I.98–99; EW I.111). However, if interested in the Earth's diurnal motion, they may consider it as a body moving with a "simple circular motion," such that all its points describe the circle that the body makes (OL IV.252).

Hobbes does not use the language of "consider as" in his Eighth Objection to the *Meditations*, but we can make sense of that Objection by seeing the perspectival principle as underlying what he says about the conception SUN. Against Descartes's claim that "there are two different ideas of the sun which I find within me" (CSM I.24; AT VII.39), one which is from the senses and the other which is

SCIENTIFIC KNOWLEDGE FROM MAKING AND THE MECHANICAL MIND   93

from "astronomical reasoning," Hobbes cryptically says "[i]t seems that there is only one idea of the sun at any one time, irrespective of whether we are looking at it with our eyes, or our reasoning gives us to understand that it is many times larger than it appears" (CSM II.129; AT VII.184). Hobbes describes the latter as a "rational inference," and I suggest that he has in mind that we can do work in astronomy only if we first consider the single conception SUN in a certain way, namely, as a body responsible for certain motions (*lux*) that are propagated through media and result in our perception of light (*lumen*) (discussed further in chapter 5). Hobbes admits that there may be "different ideas of the sun at different times" (CSM II.129; AT VII.184), such as if we look at the sun through a telescope after staring at it with our naked eye, but his point is that these are simply different images. The belief that the actual sun is larger than our conception SUN from sense experience, whether with the naked eye or a telescope, is not formed by a different idea but rather by considering SUN in a certain way. Seeing the perspectival principle as underlying Hobbes's view makes sense of his comment to close the brief Eighth Objection: "[. . .] astronomical arguments do not make the idea of the sun larger or smaller; they simply show that the idea that is acquired from the senses is deceptive" (CSM II.129; AT VII.184). Given our interests and aims, we consider SUN in a way that enables us to do astronomy; we do not "enlarge" or have a completely different conception of SUN. Hobbes and Descartes are clearly speaking past each other, since Descartes retorts that "what the objector says is not an idea of the sun [. . .] is precisely what I call an idea" (CSM II.130; AT VII.184). Hobbes's view—though not made explicit in the Eighth Objection—is that we can consider our conceptions in many ways depending on our interests.

Returning to the case of the conception EARTH, the external body under scrutiny is the same *body* but the perspectival principle informs both the way the conception of that body is considered as well as which geometrical principle is employed. The aims and interests of the natural philosopher determine the explanandum, whether diurnal or annual motion, and the *explanadum* determines the *explanans* (whether considering EARTH as POINT and then using geometry principles related to that consideration or considering as BODY with simple circular motion and then using the principles related to that consideration).

3.7 Conclusion

This chapter has shown how Hobbesian *scientia* differs from sense-based *cognitio*. Rather than having its root in many accumulated conceptions from experiences, as *cognitio* does, Hobbesian *scientia* requires the construction of a particular which serves as the foundation of an inference from that one to all. Hobbes explains this behavior by positing that humans have a unique drive that sets them apart from other animals—the passion of curiosity—and that some of those humans have the requisite leisure. Hobbes saw himself as providing the third element: method. Unlike the claims of many scholars, my account shows a diminished, though necessary, role for language insofar as it allows knowers to mark their conceptual discoveries. Throughout, the aims and interests of knowers inform this activity, both in terms of when an analysis is terminated and in terms of how instances of *scientia* are employed.

The primary goal of this chapter has been to provide an interpretation of Hobbes's view of *scientia* that provided him with the foundation for what he saw as his novel contribution to history. In the Epistle Dedicatory of *De corpore*, he asserts that, alongside the likes of Copernicus, Galileo, and Harvey, he would be remembered as having inaugurated a new science of civil philosophy. The only way for civil philosophy to be scientific, Hobbes held, was to ground it in a knowledge of causes, and he held that causes were accessible only to those who acted as makers. Hobbes's epistemology that grounded his civil philosophy was thus an amalgamation of insights from thinking about nature mechanically, a belief that knowledge is only through knowing the causes (something he found in Aristotle), and a conviction that truly knowing how something is made is possible only when one has made it. Insofar as Hobbes viewed philosophy as like "corn" and "wine"— found in nature but needing cultivation—the insight that someone knows something only if she has made it is an intuitive, commonsense idea. Artisans of all sorts would endorse this intuitive claim at the foundation of Hobbesian philosophy—no matter how many instruction booklets one reads, one fails to have knowledge unless one makes something.

Stepping back from the interpretation of Hobbes's thought, we can ask whether his principle of maker's knowledge should be taken seriously at all when his resulting geometry on which it was founded was an utter failure, most significantly obvious in his many attempts to square the circle and his recalcitrance in the face of defeat (Jesseph 1999 details these attempts and the many criticisms Hobbes faced). Despite Hobbes's failed geometry, I suggest that his emphasis on *scientia* as possible only for makers is an aspect of his view that is worthwhile and should be better known, even if only to place it within its broader historical context. As this chapter began, Hobbes's claim that knowing with certainty is possible only by doing something—making—makes it a special form of knowing-how. Although there have been some recent attempts to show that knowing-how is reducible to knowing-that (especially Stanley and Williamson 2001), there has been significant resistance. Hobbes's understanding of *scientia*, grounded in the view that there is something epistemically basic in knowledge one possesses by engaging in some skilled activity, agrees with such resistance.

4

Demonstrating Scientific Knowledge in Geometry and Civil Philosophy

How to understand Hobbes's claim that civil philosophy is a science has been a continual point of disagreement among scholars.[1] One well-known approach aligns civil philosophy's method with the *more geometrico* and sees Hobbes as fashioning it according to the model of a Euclidean demonstration. According to this "definitivist" understanding of the Laws of Nature, one begins with an axiom, say, the definition of a Law of Nature in *Leviathan* 14, and from that axiom derives the Laws of Nature found in chapters 14 and 15.[2] However, the definitivist account is difficult to hold alongside Hobbes's views about geometry and his criticisms of Euclid. For example, as discussed in chapter 3, since Euclid's definitions are not generative, Hobbes argues that "they ought not to be numbered among the principles of geometry" (EW VII.184). Furthermore, the foundation of Hobbes's own geometry is not axiomatic at all, since Hobbes holds that even Euclidean axioms need to be demonstrated (OL I.72; EW I.82). Instead, Hobbes grounds geometry in what he calls definitions by "explication" of simplest conceptions such as BODY and MOTION.[3]

In this chapter, I argue that the demonstrations that Hobbes provides in geometry and civil philosophy are instances of Making Type 3 (see section 3.4 in chapter 3). In these disciplines, teachers impart maker's knowledge to students through demonstrations. With a demonstration one moves beyond a private mark and shows another how to make something using signs. The motivation for connecting these two sciences is straightforward on Hobbesian grounds since, as has already been shown, according to Hobbes these are both instances of *scientia*; indeed, they are the only instances of Hobbesian *scientia* in the stricter sense of knowing that was the focus of chapter 3, i.e., disciplines in which knowledge

---

[1] An earlier version of parts of this chapter appeared as Adams (2019).

[2] See Deigh (1996) for this view. For discussion and criticism, see Deigh (2003), Hoekstra (2003), and Murphy (2000). See also Lloyd (2009, 151–210) for discussion of the definitivist, or as Lloyd calls it the "definitional," view.

[3] Hattab (2014, 477–478) agrees that Hobbes's first principles are not axioms but are instead definitions of simplest conceptions; however, Hattab claims that definitions of simples are "stipulative." Instead of stipulation, I argue that Hobbes intends thought experiments to provide explications of simplest conceptions already possessed from sense experience and that Hobbes holds that these can be correct or incorrect.

Hobbes's Two Sciences. Marcus P. Adams, Oxford University Press. © Marcus P. Adams 2025.
DOI: 10.1093/9780198924715.003.0004

of actual causes is possible. I argue that demonstrations in both are founded in the same manner insofar as each begins with a thought experiment in which, by undergoing the experiment, one gains the principles of that science by means of definitions by explication of what Hobbes calls "simplest conceptions." In other words, demonstrations in civil philosophy and geometry follow the same procedure. Furthermore, I have distinguished the discussion of "discovery" in the previous chapter from demonstration since Hobbes himself does so when in *Leviathan* he describes the second "speciall use" of language: we show others "that knowledge [1668: *scientiam*]" (2012, 50; 1651, 13) we have discovered by using public signs. I argue that this second special use is when a teacher employs language to advise a student how to "consider as" what is being constructed in a demonstration.

Section 4.1 connects thought experiments to Hobbes's discussions of definitions by "explication" and synthetic demonstrations. Section 4.2 examines what I call the annihilation of the world thought experiment and the state of nature thought experiment and shows how they provide definitions by explication for the simple conceptions in each domain. In section 4.3, I argue that these definitions by explication are then used as the component parts—the "causes"—of constructions in civil philosophy and geometry. Section 4.4 shows benefits of this understanding of the Laws of Nature over competing interpretations by Deigh (1996; 2003), Martinich (2002), and Lloyd (2009). Section 4.5 outlines how this interpretation impacts our understanding of the relationship of the first two parts of *Leviathan* to one another by suggesting that *Leviathan* Part II plays the role of corollaries in geometry insofar as Hobbes shows properties which follow necessarily from COMMONWEALTH but which are not contained in its essence. Ultimately, I show that the explicated definitions from the state of nature thought experiment are the beginning of Hobbes's synthetic demonstration that the Laws of Nature cause PEACE. In just the same way that Hobbes says that the geometer should "put together" the parts of a SQUARE to learn its cause, I argue that the Laws of Nature, considered jointly, are the cause of PEACE.

4.1 Demonstration and Explication: Using a Thought Experiment to Ground a Causal Science

Recall that Hobbes distinguishes two paths of philosophizing in *De corpore* I.2 and elsewhere: "Philosophy is knowledge acquired through proper ratiocination of effects or phenomena from conceptions of their causes or generations, and again of generations which are possible, from known effects" (OL I.2; EW I.3). Geometry and civil philosophy follow the first route, but how do knowers *share* with someone else knowledge of the actual "causes or generations" that they themselves have discovered in these sciences? How would Hobbes, the putative discoverer of the cause of PEACE, demonstrate how to bring it about to the readers of *Leviathan*? In this section, I argue that Hobbes could have provided an analysis aimed at establishing the first principles (simplest conceptions) of these two *scientiae* but instead, according to his method, offers a synthetic demonstration that begins in a thought experiment.

As discussed in chapter 3 with Hobbes's example of SQUARE from *De corpore* VI.6, Hobbesian analysis for discovery begins with a unified, conglomerate conception and breaks it apart into simpler conceptions, ultimately ending in the conceptions that are simplest for that domain such as BODY and MOTION. As discussed below, Hobbes talks about such analysis as proceeding "from the sense-experience of things to universal principles" (Hobbes 1981, 313; OL I.71; EW I.80)—the process of resolving or subtracting. If Hobbes had moved by analysis to first principles and from first principles by synthesis in the civil philosophy of *Leviathan*, he would have begun that work by analyzing a peaceful commonwealth into the Laws of Nature and the bodies of that domain (humans) and their motions (passions). Then, using those bodies and their motions as principles, he would have synthetically re-constructed the peaceful commonwealth (this would be a case of Making Type 1, described in section 3.4 in chapter 3).

Anyone familiar with *Leviathan* knows that this is not at all how Hobbes proceeds. Instead of such an analysis, Hobbes employs a thought experiment to establish first principles because he holds that in a demonstration one should provide only a synthesis (OL I.71; EW I.80–81). I suggest that Hobbesian thought experiments thus function as a shortcut to the conceptual bedrock for the domain of interest, taking the place of an analysis so that demonstration can proceed only by synthesis. But how can we be said to be moving from "known" causes, as the first route of philosophizing mentioned above, if we require a thought experiment? Insofar as these known causes (simple conceptions) are contained as 'parts' in every conception received in sense experience, they are known to anyone with any sort of experience. Indeed, Hobbes claims in *De corpore* VI.2 that

> [. . .] the causes of the parts are better known than the causes of the whole. For the cause of the whole consists of the causes of the parts; and it is necessary that the components be known earlier than the composite. But by 'parts' I understand in this place not the parts of the thing itself but the parts of its nature, so that by the parts of a man, I do not understand head, shoulders, arms, and so on, but figure, quantity, motion, sensation, reasoning, and similar things, which are accidents which assembled at the same time constitute the whole man—not his bulk but his nature. (Hobbes 1981, 291; OL I.60; EW I.67)

The purpose of a thought experiment is to "explicate" (Hobbes's term, discussed more below) these simples, such as MOTION, that experienced individuals already know. We never experience simple conceptions *qua* simples (see fn. 10), but we find them contained in unified, conglomerate conceptions. Working through a thought experiment clarifies these conceptions (and registers that clarity with a definition by explication) as well as the relationships *among* conceptions, such as the

98   HOBBES'S TWO SCIENCES

relationships between TIME and MOTION and between HOPE and EQUALITY
(discussed below in subsections 4.2.1 and 4.2.2). The remainder of this section will
outline Hobbes's account of demonstration and definition by explication.

According to Hobbes, a demonstration shows the process of moving from principles to an end.[4] He states in *De corpore* VI.16 that "a demonstration is a syllogism
or series of syllogisms from the definitions of names all the way to the final conclusion" (Hobbes 1981, 323; OL I.76; EW I.86).[5] Hobbes connects this understanding
to the "origin of the name" and "showing": "[. . .] even though the Greeks used
ἀποδείξιν which the Latins translated as the word *demonstratio*, only for that reasoning in which they, having described certain lines and figures, put the thing practically before their eyes, which properly is ἀποδεικνύειν or *monstrare*" (Hobbes
1981, 323; OL I.76; EW I.86). This definition makes clear that demonstrations are
possible only for those capable of engaging in syllogistic reasoning and *a fortiori*
only for those who know the proper definitions of names from which to begin a
syllogism. However, this definition does not say precisely how language aids
demonstration.

As discussed in chapter 2, Hobbes holds that the mental activities of adding and
subtracting can occur independently from language (Hobbes 1981, 179; OL I.4;
EW I.3–4). Hobbes later states the "entire method of demonstrating is synthetic"
(OL I.71; EW I.80–81) and elsewhere describes synthesis as "ratiocination from
the first causes of construction through the middle [causes] continued to the very
thing itself" (OL I.254; EW I.312).[6] Hobbes's use of 'ratiocination' here and in his
definition of philosophy at *De corpore* I.2 links synthesis with the general mental
activity of adding and subtracting: "And by ratiocination I understand computation [*computationem*]. To compute truly is to collect the sum of many things added
together at once, or when one thing from another subtracted, to know the surplus"
(OL I.3; EW I.3).

Each syllogism in the series of syllogisms making up a demonstration will be
composed of definitions or of conclusions already derived from definitions (OL
I.76; EW I.86). In *De corpore* IV.8, Hobbes offers the following syllogism when
describing what corresponds in the mind to syllogistic reasoning:

> Man is an animal;
> An animal is a body;
> Therefore, man is a body.

---

[4] The term 'demonstration', Hobbes argues, should be reserved only for the reason 'why' (OL IV.38).
For discussion, see Jesseph (1999, 204–205).

[5] I have modified Martinich's translation (Hobbes 1981) here and in quotations that follow this,
using 'demonstration' rather than 'proof' for *demonstratio*.

[6] More broadly, Talaska (1988, 214) has argued that Parts I–III of *De corpore* are a synthetic
demonstration.

The mind evaluates the truth of the premises such as these, as well as the connection between the premises and the conclusion, by means of definitions. For example, he argues that we evaluate the truth of the conclusion by imagining a man and conceiving that individual as in some place or occupying space (a particular considered as universal). We then remember that whatever appears as such, by the definition of 'body' (discussed below), is called by the name 'body' (OL I.44; EW I.50).

This method of evaluating a syllogism suggests that language used in a demonstration is a tool to aid in examining conceptions and the connections conceived among them. Hobbes makes this role of language evident in *De corpore* IV.8 when describing "[w]hat in the mind might answer to a syllogism." There he reduces knowing that a conclusion of a syllogism is true to recognizing that the *conceptions* signified by its names are necessarily connected to one another in a unified, conglomerate conception:[7]

> The thinking in the mind answering to the direct syllogism is like this: first a phantasm of the thing named with the accident or attribute of it, on account of which it is called by the name which is the subject in the minor proposition, is conceived [*concipitur*]; then a phantasm of the same thing with the accident or attribute on account of which it is called by name which is the predicate in the same proposition. Third, the thinking [*cogitatio*] returns again to the thing named, with the attribute on account of which it is called by the name which is in the predicate of the major proposition. Finally, when a person remembers that those attributes are all one and the same thing, he concludes that those three names are also names of the same thing; that is that the conclusion is true. (Hobbes 1981, 259; OL I.44; EW I.49–50)

Teachers demonstrate to students using names and propositions to aid this mental process, but at its core, demonstration requires conceiving connections among *conceptions*. Remembering "that those attributes are all one and the same thing" with language, as Hobbes describes above, is just to recall linkages discovered among conceptions that are received in sense and considered in various ways. Ultimately, I argue in section 4.3 below, this is how the demonstration of the Laws of Nature counts as a syllogistic (synthetic) demonstration: when all of the Laws of Nature are simultaneously performed, they jointly cause PEACE.

---

[7] For discussion, see Pécharman (2016, 52–53). Pécharman agrees that in *De corpore* the act of *conceiving* accidents as belonging to the same subject (connections conceived among conceptions) is the cause of a conclusion in language being true and, furthermore, argues that by reducing syllogistic reasoning to computation Hobbes has made the "traditional middle term obsolete."

100   HOBBES'S TWO SCIENCES

What can Hobbes mean in claiming that a demonstration, understood as showing, will be synthetic only (OL I.71; EW I.80–81)? Hobbes views demonstration as a form of teaching (*docere*) where the teacher shows the student the "tracks" by which the teacher discovered a conclusion. He claims that the

> [. . .] method of demonstration will be the same as the method of discovery had been, except that the first part of the method, namely, that which proceeded from the sense-experience of things to universal principles, is to be omitted. For the latter things cannot be demonstrated, since they are principles; and as they are known by nature (as was said above in section 5), they do indeed need *explication*, but not demonstration. (Hobbes 1981, 313; OL I.71; EW I.80–81; emphasis added)

Later, in *De corpore* 20.6, Hobbes claims that "*Synthetica* is the art itself of demonstration" (OL I.252; EW I.310). The untaught individual, like Adam, without a teacher to demonstrate might discover by means of analysis from sense experience to principles, but this analysis should be omitted when Adam as a teacher offers a demonstration to a Cain or an Abel. Instead of a demonstration for the principles of a synthesis, an explication is needed.

It seems that Hobbes means something close to the etymology of 'explicate', something like an unfolding but related to what we do with conceptions when we inspect their contents. Indeed, Hobbes uses a similar notion later in *De corpore* VI.14 when he gives the definition of 'definition'. There he defines 'definition' as "a proposition, the predicate of which is the resolution of the subject, when it can be done, and is its *explication* when it cannot be done" (Hobbes 1981, 319; OL I.74; EW I.83–84; emphasis added).[8] When defining by resolution is impossible, we reach a definitional regress stopping point by explicating the word in question. Since Hobbesian names used properly signify our conceptions, lest we commit an absurdity, we explicate by unfolding the contents of conceptions received in experience.

It will be useful to mention some of Hobbes's examples of explication and then I will provide a simple example to motivate how this might work in a demonstration. Hobbes identifies the "highest causes" in *De corpore* VI.5 and provides some examples in the following article. He claims that "the causes of universals (those of which there are any causes at all) are obvious *per se* or known to nature, as they say" (Hobbes 1981, 295; OL I.62; EW I.69). Examples that Hobbes provides in this discussion include BODY, PLACE, and MOTION, and these can serve as principles

---

[8] The *Latin Works* edition (OL I.74) follows the 1650 edition and uses *exemplicativum* here, which Martinich (in Hobbes 1981, 319) renders as 'explicate'. Schuhmann (in Hobbes 1999, 68) corrects the misspelling in *De corpore* from *exemplicativum* to *exemplificativum*, reflecting the spelling that Hobbes provides in a quotation of this passage in *Examinatio et Emendatio* (cf. OL IV.38).

for a demonstration when we have their explications: "When, therefore, universals and their causes (which are the first principles of knowledge τοῦ διότι) are known, we have first their definitions (which are nothing other than the explications of our simplest conceptions)" (Hobbes 1981, 295; OL I.62; EW I.70).[9]

What will count as a definition by explication for a simplest conception? Hobbes provides the following example of PLACE: "[...] whoever *rightly conceives* [*recte concipit*] what a place is [...] must know the definition: that a place is a space which is completely filled or occupied by a body" (Hobbes 1981, 295; OL I.62; EW I.70; emphasis added).[10] Sometimes a definition by explication will be merely stipulative, such as if someone were to define 'bachelor' as "unmarried male adult human." If one is writing a legal code that makes use of that name but fails to explicate it, readers will be unable to judge the reasoning that the legal code employs and, furthermore, be unable to apply the code consistently. In *Leviathan* 5, Hobbes indicts much of philosophy as failing to provide definitions by explication of just this sort:

> For it is most true that Cicero sayth of them somewhere; that there can be nothing so absurd, but may be found in the books of Philosophers. And the reason is manifest. For there is not one of them that begins his ratiocination from the Definitions, or Explications [1668: *Explicationibus*] of the names they are to use; which is a method that hath been used onely in Geometry; whose Conclusions have thereby been made indisputable. (2012, 68; 1651, 20)

The first cause of absurdity that Hobbes identifies is a failure to begin ratiocination "from the settled signification of [...] words" (2012, 70; 1651, 20). Beginning a legal code by explicating 'bachelor', and other names used, makes it possible to settle whether and, if so, how each of those names signifies conceptions. So far, it seems that this might be a completely arbitrary task. For example, if I plan to use a proposition like "All plinks are plunks" in a contract, then I may declare that I will use the name 'plink' to signify "plunks I have seen in Plinkville" and then also could

---

[9] Hobbes seems to use 'cause' in this quotation in the following way: when we have the correct definitions of "our simplest conceptions", we then know the cause of those very conceptions. For example, when we know that "a place is a space which is completely filled or occupied by a body" (Hobbes 1981, 295; OL I.62; EW I.70), we know what causes the simplest conception PLACE, namely, bodies filling space. For an argument that Hobbes's philosophical method aims at knowledge of the causes of our *conceptions* of things, not at the cause of things themselves, see Hattab (2014, 474).

[10] Boonin-Vail (1994, 32) argues that definitions of these simples, in this case the definition of PLACE, are propositions "whose truth depends upon [their] correspondence to reality," but Hobbes could not hold that we evaluate them in this way. We would not look simply to "reality," as it were, to judge this definition, since we will never encounter PLACE *qua* PLACE in experience. Instead, we discover the conception PLACE as it is contained within each of our conceptions of individual bodies. Thus the one who "correctly conceives" the definition of 'place' can do so only by either an explication by means of a thought experiment or an analysis from a conception of a particular body (discussed below).

explicate 'plunk'. Having explicated these names by mere stipulation, I can ratiocinate from that beginning, and those to whom I utter "All plinks are plunks" will be able to evaluate my conclusions.

Definitions by explication of *simplest conceptions*, however, are not stipulative according to Hobbes. As mentioned already, Hobbes notes that having the right definition by explication of 'place' depends upon "rightly" (*recte*) conceiving PLACE. Recognizing this difference between stipulative and non-stipulative Hobbesian explicatory definitions—that simplest conceptions from sense experience are either rightly or wrongly conceived—gives reason to resist understanding Hobbesian natural philosophy as a type of conventionalism, according to which natural philosophy is founded on stipulative, arbitrary definitions.[11] Indeed, although simplest conceptions are manifest *per se*, Hobbes does not hold that every person has in fact correctly conceived BODY or PLACE—*mis*conception is possible. For example, he argues in *Anti-White* III.5 that Aristotle's definition of 'place' is straightforwardly "not true [*vera non est*]" (Hobbes 1973, 120). The issue is not that Aristotle failed to provide a clear definition but rather that the names used fail to signify the correct manner of conceiving the simplest conception PLACE.

Similarly, in *Six Lessons* Hobbes argues against defining 'point' as "having no quantity at all." Such a definition is not an unclear stipulatory definition, rather it fails to provide what is needed for later demonstrations because it signifies a *mis*conception; POINT should rightly be conceived as a body for which no magnitude is considered. As Hobbes says, "[i]f a point have no quantity, a line can have no latitude," and thus it would be impossible ever to draw a straight line (EW VII.219). Clarity in definitions by explication (whether stipulative or not) is necessary but not sufficient: "[. . .] their [i.e., definitions'] use is, when they are *truly* and *clearly* made, to draw arguments from them and for their conclusions to be proved" (EW VII.220; emphasis added). In other words, no matter how clear one's stipulative explicatory definition may be, whether of 'point' or 'place' or otherwise, it will also need to be based upon rightly conceiving one's conceptions to be of any value for *scientia*.

Hobbes distinguishes between what I am identifying as non-stipulative, explicatory definitions and generative definitions earlier in *Six Lessons* by alluding to the traditional distinction between definitions, postulates ("petitions"), and axioms ("common notions"). He notes that the principles of demonstration are definitions (something like the explicatory definition for 'point'), but he contrasts these explicatory definitions with petitions by calling the latter "principles of construction," such as *"that a man can draw a straight line, and produce it"* (EW VII.199). His point is that one must first begin with a postulate related to 'straight line' before being able to construct more complex figures like squares, so he calls these

---

[11] For more details regarding the conventionalist view, see chapter 6, fn. 3.

DEMONSTRATING SCIENTIFIC KNOWLEDGE   103

"principles of operation" (EW VII.199). However, the only way to understand 'straight line' correctly is to have the proper explicatory definitions for 'point' and 'motion' already in place.

Hobbes later applies this distinction between explicatory and generative definitions to diagnose (and in his view solve) a longstanding disagreement among geometers concerning the angle of contact.[12] Hobbes's treatment of the angle of contact shows the centrality of this distinction. The angle of contact is the angle formed between a circle and its tangent, and commentators since Euclid had worried that Euclid's account of it seemed inconsistent with Proposition I of Book 10 related to unequal magnitudes. Without discussing the details of this history,[13] these worries arose from Euclid's accounts of the different types of angles and the relationships between unequal magnitudes: on the one hand, Euclid showed that no angle of contact (also called a "horn angle") could be greater than any acute rectilinear angle (Proposition 16 of Book 3), but on the other hand, as Jesseph notes (1999, 161), angles of contact can be organized from greater to lesser magnitudes according to the size of the radius of the circle on which they are created. Given the latter consideration, if a horn angle were of a lesser magnitude than an acute rectilinear angle then, given Proposition I of Book 10, there *should* be some magnitude of it that, after subtraction, would exceed an acute rectilinear angle that was initially of greater magnitude. But that result should not be possible given Euclid's demonstration (Proposition 16 of Book 3) that no horn angle could exceed the magnitude of any acute rectilinear angle.

Commentators tried different strategies to avoid holding that Euclid's account of the angle of contact was inconsistent, including denying that angles were quantities. Wallis's attempted solution involved providing a definition by explication for 'angle', but Hobbes argued that this approach failed to solve the worries. As Jesseph (1999, 167–172) details, Hobbes describes his own solution in *De corpore* XIV.7 and distinguishes angles of contact as being constructed differently from rectilinear angles. While "angles of circumlation" (rectilinear angles) are made by rotating a line around one of its endpoints, "angles of contingence" (angles of contact) are made by a different type of motion that Hobbes describes as "continual flexion or curvation in every imaginable point" (OL I.160; EW I.184; see also *Examinatio*, Dialogue V, OL IV.161ff). This strategy of defining different types of angles with generative definitions allows Hobbes to dissolve the apparent inconsistency in Euclid's account (and Hobbes thinks Wallis's) since, on the one hand, both the rectilinear angles and angles of contacts can be understood as quantities,

---

[12] I thank an anonymous reviewer of this press for emphasizing the importance of Hobbes's views on the angle of contact.
[13] For these details and more on Hobbes's own account, see Jesseph (1999, 99–100, 159–172; also Jesseph 1993).

104 HOBBES'S TWO SCIENCES

but, on the other hand, they are formed by different types of motions and are thus subject to different types of measurement.

Returning to the explicatory definitions of simplest conceptions, it should be clear that if one misconceives a conception like PLACE, like Hobbes thinks Aristotle does, then the definition one uses to signify that misconception will be in error. If Hobbes held that explicatory definitions that serve as the principles undergirding constructions in geometry are either correct or not, and thus more than mere stipulation, how do we avoid misconception and arrive at the right explication? Part of the answer to this question is reflected in one Hobbes's broader projects in *Leviathan*: showing how to avoid absurdity by "meditating" upon one's conceptions to be sure that each name used signifies a conception received in sense experience.[14] For example, in *Leviathan* 13 Hobbes argues that knowers who pay attention to their own conceptions, "by their own meditation, arrive to the acknowledgment of one Infinite, Omnipotent, and Eternall God, chose rather to confess he is Incomprehensible" (Hobbes 2012, 168; 1651, 53). When we examine our conceptions carefully, we discover that we lack any such conception for God as infinite, omnipotent, and eternal and so declare that a definition of God's nature as "*Spirit Incorporeall*" is "unintelligible" (Hobbes 2012, 168; 1651, 53).[15] He furthermore argues in *Leviathan* 3 that we avoid "absurd speeches" when we recognize that "the Name of God is used, not to make us conceive him; (for he is *Incomprehensible*; and his greatnesse, and power are unconceivable;) but that we may honour him" (Hobbes 2012, 46; 1651, 11).

Avoiding absurdity thus requires (1) checking whether a name used signifies a conception and (2) registering the correct way of considering a conception. Most of the time, we define non-absurd names by resolving them into their component parts. We resolve, for example, the name 'man' into the names 'body', 'animated', and 'rational' (OL I.4; EW I.4). Defining 'man' as "animated rational body" corresponds to linkages that are present at the conceptual level for MAN, which linkages exist even apart from us having names with which to signify them (i.e., even if "no words were imposed"; cf. *De corpore* 1.3; Hobbes 1981, 176–177; OL I.3; EW I.3–4). In crafting definitions, care must be taken so that each component part (names), as well as the linkages made between them, signifies a conception received in experience, or contained in one that is, as well as conceptual connections received in experience. For example, if we define 'man' as "animated rational flying winged body" we have not used any absurd *names*, since each name considered singly signifies conceptions received in experience; however,

---

[14] See the discussion of 'meditate' and 'contemplate' in chapter 3 (fns. 46 and 49).

[15] Similarly, Hobbes holds in *Elements* that although "we can have no conception or image of the *Deity* [...]" we do have a conception "that *there is a God*" (EL I.XI.2; 1650, 132). This conception is one of God as first cause, and any time that we attempt to ascribe attributes to God, the names that we use, such as 'omnipotent', "*signifie* [only] either *our incapacity*, or our *reverence*" (EL I.XI.3; 1650, 134).

we have made a linkage between names that is not present in conceptual linkages from experience.

If we need to resolve 'man' further than "animated rational body," we may do so, but eventually we will reach a point where resolution is no longer possible because we have reached a name that signifies a simple conception. These simples, such as BODY, are non-absurd, since each will be contained in conceptions received in experience, but we must provide an explication for each of them that we use. Thought experiments for Hobbes help us gain such an explication whereby we can know that we have not misconceived, and these simple conceptions can then serve as the starting points for a demonstration by synthesis.

A simple example will help distinguish the different roles played by conceptions and language in demonstrating, in particular the role of language in definitions by explication. Imagine that I aim to teach a student how to make a circle so that they can determine its properties (Hobbes 1981, 181; OL I.5; EW I.6). I would demonstrate this not by mere verbal instructions but would do so by putting a particular point into motion using a compass to hold equal all the radii as I construct. Insofar as the student themselves attends to the action of moving the compass as they draw their own circle, holding it steadfast through the entire motion of the line that forms the circumference of the circle, they seem to possess maker's knowledge (*scientia*). They know *that* object's causes, since they themselves were its maker; showing them how to make it required no analysis. However, imagine that in teaching my student how to draw a circle I provided multiple demonstrations of myself making circles, some with standard graphite pencils and others with colored pencils or markers. Imagine that I also varied the surfaces on which I drew, drawing some circles on a chalkboard and others on construction paper. If after demonstrating these instances of circle making to my student, the student then produced only circles with a purple outline, it would seem that the student missed something. This shortcoming would be more evident if, when there were no purple-colored pencils available, the student would refuse a request to make a circle, saying that they were not able to make a circle without *that* colored pencil. Maker's knowledge of only this or that purple circle would fail the usefulness test for *scientia* discussed in chapter 3: "[k]nowledge is for the sake of power [*Scientia propter potentiam*]" (Hobbes 1981, 183; OL I.6; EW I.7).

Clearly, the student described above knows how to make a circle, that is, a purple-outlined circle, but their causal knowledge (*scientia*) is severely limited. They have mistakenly latched onto the use of that type of pencil and see its use as part of what is required to make the thing they watched me make. This deficiency in the student's grasp illuminates where considering as arises for maker's knowledge, for it is not just knowing how to make *this* or *that* circle that Hobbes attributes to *scientia*. Instead, it requires the same cognitive activity that enabled the analysis described in chapter 3, wherein I considered some particular CLOSET DOOR as SQUARE. In starting with something shown to them by a teacher in demonstration (a synthesis that omits the teacher's own analysis), the student

106 HOBBES'S TWO SCIENCES

who gains *scientia* of CIRCLE must consider their own pencil as an instantaneous POINT and then consider the movement of the compass as MOTION. For this to occur, the teacher must supply definitions by explication, such as "a body is [. . .]" or "motion is [. . .]", so that the teacher can next provide definitions such as "a point is a body considered without magnitude" to instruct the student how to consider the pencil as it moves with the compass. Additionally, the student must consider the feature EQUAL as holding throughout the construction and be told, using language, to ignore irrelevant features of their conceptions, such as the particular lengths of each radii, say, 2 inch radii, and the color of the pencil used in drawing it. Ignoring what is irrelevant was crucial to the account of the triangle, as seen in the discussion in chapter 3, and the discovery of a universal rule (2012, 54; 1651, 14), but in the case of demonstrating language functions to tell another person precisely *what* to ignore and *what* to attend to, i.e., how to "consider as" that which the teacher is showing them through the demonstration. This begins with definitions by explication and continues throughout the demonstration as the teacher provides more definitions.

In the next section, I discuss two Hobbesian thought experiments that I argue provide definitions by explication to do just this: the annihilation of the world thought experiment in *De corpore* VII–VIII and the state of nature thought experiment in *Leviathan*. For the sake of clarity in the connection that I hope to make between the two thought experiments, I will identify three steps that Hobbes takes in these thought experiments: in step 1, Hobbes asks the reader to imagine a scenario by privation; in step 2, the reader's attention is drawn to features of the bodies within the imagined scenario; and in step 3, Hobbes adds something to the imagined scenario and considers what follows from its addition. In both thought experiments, Hobbes sought to aid readers in rightly conceiving their conceptions, not simply stipulating explicatory definitions.

4.2 Hobbes's Two Thought Experiments

4.2.1 The Annihilation of the World Thought Experiment to Ground Geometry

Part II of *De corpore* is entitled "The First Grounds of Philosophy." Hobbes understands the function of first philosophy as providing a theory of body by means of our phantasms and not of being *qua* being (Leijenhorst 2002, 101). Thus, Hobbes begins *De corpore* VII.1 by identifying his goal of providing an account of the "teaching of natural philosophy." Hobbes claims that first philosophy of this sort, which will serve as the foundation for geometry and, ultimately, for natural philosophy, begins best from "privation" (I describe another role that first philosophy plays within natural philosophy in chapter 5 when considering Hobbes's two *a priori* principles concerning motion and rest in *De corpore* VIII.19).[16] The privation to be considered is a situation in which the entire universe has been destroyed save for one lone individual (step 1). Implicit in this starting point is that the world that caused the conceptions that remain in the man's mind did exist at one time (Sorell 1995, 92).

Having been asked to consider such a scenario, the reader examines the features of the body in that system (step 2). We are first asked what would be available to ratiocinate upon for the lone individual who remained. Hobbes asserts that all the "ideas," or phantasms, of the things that this individual had seen or otherwise perceived would remain for him: "Therefore, to these [phantasms] he would impose names, these he would subtract and compound" (OL I.82; EW I.92).[17] This lone individual would be in a position similar to everyday experience, for when examining the world around us "we compute nothing but our phantasms" (OL I.82; EW I.92). Astronomy is a case in point: when we measure motion in astronomy, we do not ascend into the heavens; instead, "we do so quietly in our closet or in the dark" (OL I.82; EW I.92). We use names that signify phantasms, but we are really examining phantasms, and connections between them. Although we may use signs in first philosophy, such as 'body', those signs are shared memorials that signify conceptions, such as BODY. The difference between this remaining individual and those in the everyday world is that he has only ideas from past experiences.

Regardless of which conception this individual examined, first he would find contained within it the conception SPACE. This conception would be a "mere phantasm" (OL I.82; EW I.93), since it would not, when considered, be caused by an external body. As a result, this conception is "imaginary" SPACE, defined as "the phantasm of an existing thing, so far as it is existing, i.e., when no other accident of a thing is considered except that it appears outside of imagining" (OL I.82; EW I.94).[18] In other words, imaginary SPACE is the phantasm that arises when considering the conglomerate conception of any body whatsoever and finding contained within that conception the conception that the body that caused it existed outside

---

[16] Hobbes uses a brief thought experiment by "privation" in *Dialogus Physicus* (1661, 1668) when speaker A articulates the view that since simple circular motion is congenital to the Earth it is also congenital to each "one of its atoms" (OL IV.253). Speaker A supports this claim by asking what would result if the Earth were "annihilated by divine omnipotence or if half this Earth were removed to some other distant place beyond the fixed stars" (see discussion in Adams 2017, 91). For additional privational thought experiments that Hobbes offers, see *Anti-White* III.1 (Hobbes 1973, 117) and *Elements* (EL I.I.7). Leijenhorst (2002, 110, fn. 49) sees the annihilation thought experiment in *De corpore* VII–VIII as an "argumentative device" wherein we *suppose* that annihilation has happened, which as a mere supposition Hobbes can consistently hold alongside his claim in *Anti-White* XIII.8 that "the universe or a part of it cannot perish, unless it has been supernaturally annihilated by God" (Hobbes 1973, 193; 1976, 145) and in *Dialogus Physicus* speaker A's reference to "divine omnipotence."

[17] Hobbes uses 'conception', 'phantasm', and, to a lesser extent, 'idea' interchangeably; my usage reflects this.

[18] For discussion of Hobbes on imaginary space, see Jesseph (2016, 136; 2015) and Slowik (2014; 2021).

108 HOBBES'S TWO SCIENCES

of the perceiver's imagination. Similarly, Hobbes identifies imaginary SPACE as the "image, or phantasm of a body" in *Anti-White* III.1 (Hobbes 1973, 117).

The next after-effect of past experiences discovered is TIME. Just like "a body leaves a phantasm of its magnitude in the mind" (OL I.83; EW I.94), resulting in imaginary SPACE, a previously-experienced moving body leaves the phantasm of TIME. As a result, when ratiocinating upon some of his conceptions, this lone individual in the thought experiment would discover MOTION contained within them.[19] MOTION just is the "idea of a body crossing, now through this, now through that space in continual succession" (OL I.83; EW I.94).[20] Hobbes is explicit that the phantasm TIME is conceptually dependent upon motion, preventing us from claiming that "time is the measure of motion." Instead, "we measure time by motion, not motion by time" (OL I.84; EW I.95). For example, we measure the passing time of a day by the motion of the shadow around the sun dial. Here it seems that the thought experiment is not merely aiding in the explication of the conception TIME but, in doing so, clarifying its relationship to other conceptions such as MOTION. We thus have discovered that bodies in motion would have left two after-effects in the mind of the lone individual: the magnitude of any body leaves the phantasm of imaginary SPACE, and the motion of some bodies leaves MOTION and from this TIME.

These definitions of imaginary SPACE, MOTION, and TIME have been explicated insofar as they were learned by examining the contents of particular conceptions. Were the lone individual to continue ratiocinating, additional definitions that would be discovered include PART, understood as a relation of one section of imaginary SPACE compared with the larger SPACE that contains it, and DIVISION, which is understood as "nothing else but to consider one and another within the same" (OL I.84; EW I.95). After these, the individual could gain ONE, NUMBER, COMPOUND, and WHOLE. Each of these conceptions is understood in terms of imaginary SPACE.

De corpore VIII.1 begins step 3 of the thought experiment as Hobbes departs from considering only the lone individual by reintroducing one body: "Let us suppose next that one of the things be put back, or created once more" (OL I.90; EW I.102). The reader discovers two things: first, this thing, whatever it may be, would

---

[19] Not all conglomerate conceptions of particular bodies will contain MOTION; some will contain REST. Hobbes acknowledges that unlike the accident of extension, without which we cannot conceive a body, there are accidents shared by some but not all bodies, including resting, being moved, color, and so on (*De corpore* VIII.3; OL I.93; EW I.104). Slowik (2014, 78) suggests that motion and rest are disjunctive properties of bodies insofar as one or the other is always possessed by a body.

[20] Were Hobbes being consistent with terminology, he might call the conception of MOTION implicated at this stage in the thought experiment *imaginary* MOTION since when considered it would not be caused by a moving body but would only be an after-effect (like imaginary SPACE). This would also imply that the conception of TIME offered at this stage of the experiment is imaginary. Such a distinction would help make sense of his later definition of (real?) MOTION in *De corpore* VIII.10, which relies upon a body being reintroduced for the lone individual to consider and the definition of PLACE that he develops (discussed below).

have no dependence upon the individual's imagination (mind independent); and second, it would be a part of imaginary SPACE such that it would be "coincident and coextended" with it (OL I.90; EW I.102) (extended). Note that we can make sense of PART only since it was defined earlier in chapter VII with reference to imaginary SPACE. Hobbes identifies the resulting clarified conception BODY with the name 'body' (OL I.91; EW I.102) and provides a definition by explication that uses these two aspects of the recreated thing.

The account of PLACE relates to the distinction Hobbes makes between imaginary SPACE and the extension, or real SPACE, of a body in *De corpore* VIII.4–5. Unlike imaginary SPACE, Hobbes argues that extension does not depend upon our thinking (OL I.93; EW I.105); instead, the magnitude of a body causes this conception. To further distinguish PLACE from real SPACE, he argues that PLACE is the part of imaginary SPACE which is coincident with the magnitude of some body. He distinguishes PLACE from MAGNITUDE by having the reader consider what happens when a body is moved; its magnitude is retained but its place is not. From this account of PLACE we define MOTION caused by the body returned to the thought experiment for step 3: "motion is a continuous abandoning of one place and acquiring of another." Recall that TIME is an after-effect of bodies' motion, but the lone individual would not yet have been able to formulate a definition of the motion of an *existing* body even so (see fn. 20). This is only possible now with the conception PLACE explicated. Oddly, unlike in the case of imaginary/real SPACE, Hobbes provides no account of real TIME after a body is reintroduced.[21]

Notice that imaginary SPACE is the fundamental conception in Hobbes's first philosophy from which he explicates more complex conceptions such as real SPACE, BODY, PLACE, and MOTION. It might seem strange to claim that imaginary SPACE is conceptually fundamental since Hobbes believes it to be caused by bodies. Why would Hobbes see this conception as foundational when it is caused by bodies, which are *ontologically* fundamental? Bodies in motion are certainly Hobbes's ontological fundamentals; however, Hobbes's aim in the experiment is to unearth—to explicate—our most simple conceptions caused by moving bodies. Hobbes's view that only bodies in motion exist in the world is an assumption for which he does not argue, as discussed in chapter 2, and his goal is to show which conceptions are caused by experiences of those bodies. Given that aim, imaginary SPACE emerges as conceptually fundamental. Subsection 4.3.1 discusses how these conceptions explicated in the experiment are put to use in geometrical definitions as we build up from BODY and MOTION considered simply to complex geometrical figures, but first I shall consider the state of nature thought experiment.

---

[21] Gorham (2014, 95ff) argues that Hobbes needs real TIME as a corollary to imaginary TIME; he suggests that just as real SPACE (extension) is an essential attribute of body, real TIME could be 'succession' as the essential attribute of motion (for argument against, see Leijenhorst 2002, 134–137).

110   HOBBES'S TWO SCIENCES

4.2.2 The State of Nature Thought Experiment to Ground Civil Philosophy

The state of nature thought experiment in *Leviathan* 13 is perhaps the most notorious aspect of Hobbes's philosophy. Ioannis Evrigenis comments that its prominence is somewhat "paradoxical" since he claims that it "falls far short of [Hobbes's] stated standards of precision" in *Leviathan* (Evrigenis 2016, 222). The standards that Evrigenis has in mind are the careful definitions that Hobbes provides in the preceding chapters, and he suggests that "it is precisely because it violated those standards that [it] [. . .] became so successful" (Evrigenis 2016, 222). However, viewed alongside the annihilation of the world experiment from first philosophy, we see that imprecision (and, by means of that imprecision, success) cannot be Hobbes's intent. Instead, Hobbes provides a thought experiment to explicate simple conceptions and then (as will be argued in section 4.3 below) proceeds synthetically from those simples as principles. In other words, the state of nature experiment enables Hobbes to omit "the first part of the method, namely, that which proceeded from the sense-experience of things to universal principles" (Hobbes 1981, 313; OL I.71; EW I.80).

Like the annihilation thought experiment, in the state of nature thought experiment the reader is first asked to imagine a situation of privation (step 1); the reader reflects upon what would result were civil society destroyed. Thus, to learn the "natural condition of Mankind," the experiment considers human bodies devoid of their artificial civil relationships. The first property of "natural" human bodies (step 2) that the reader learns is that

> Nature hath made men so equall, in faculties of body, and mind; as that though there bee found one man sometimes manifestly stronger in body, or of quicker mind then another; yet when all is reckoned together, the difference between man, and man, is not so considerable [. . .]. (Hobbes 2012, 188; 1651, 60)

Considering any of these human bodies results in discovering that EQUALITY is contained in its conception. The EQUALITY of strength and mind that Hobbes has in view is evident from the fact that "the weakest has strength enough to kill the strongest [. . .]" (Hobbes 2012, 188; 1651, 60) if they are sufficiently cunning. We learn next that "[f]rom this equality of ability, ariseth equality of hope in the attainining of our Ends" (Hobbes 2012, 190; 1651, 61).

The reader has so far learned that an accident of all natural human bodies is EQUALITY of strength and mind and that this accident causes human bodies to be moved in a certain way by the passion of HOPE. Since Hobbes considers passions as motions, or more precisely as the "interiour beginnings of voluntary motions" which are endeavors (Hobbes 2012, 78; 1651, 23), what we have learned now is how all of these bodies move. It is an assumption that all bodies in this privational scenario, that human bodies devoid of civil relationships, are moved in one way or another by PASSIONS. This may seem to be a premise from natural philosophy, but we need not take Hobbes to think of it this way. Instead, it functions as a simple conception for civil philosophy, like BODY and MOTION do in geometry (more on this below). From their EQUALITY, they also have "appetite with an opinion of attaining" (Hobbes 2012, 84; 1651, 25) what they desire, which is the passion HOPE. The relationship of EQUALITY to HOPE is similar to the relationship of MOTION to TIME in the annihilation thought experiment. As TIME is conceptually dependent upon MOTION (discussed already), HOPE is conceptually dependent upon EQUALITY. Both EQUALITY and HOPE are accidents of all of the bodies in the system being considered (the state of nature), but we only understand HOPE in terms of EQUALITY. Prior to beginning the thought experiment, Hobbes provides the reader with *Leviathan* chapters 1–11 to describe in much greater detail the extent to which each HUMAN BODY is naturally EQUAL and is moved by PASSIONS. Each human body receives conceptions from sense experiences, these conceptions remain as decayed motion leading to imaginations, memories, and experiences, leading to prudence, these conceptions are subject to deliberation, and so on. Furthermore, each of these human bodies moves in their natural state according to appetites and aversions, articulated as motions too small to measure (endeavor).

Although these may seem like natural-philosophical claims, I take them to be providing more details than is strictly necessary about the simple conceptions that are the principles of the science of civil philosophy (I return to this topic in section 4.5 below and discuss *Leviathan* 12 in chapter 6). The thought experiment in *Leviathan* 13 is capable of standing alone for the reader who begins with it for the demonstration, and in this way, it has the same structure as the annihilation of the world thought experiment at the beginning of first philosophy. In other words, the claim "when all is reckoned together, the difference between man, and man, is not so considerable [. . .]" (Hobbes 2012, 188; 1651, 60) is sufficient for explicating EQUAL, but an interested reader may desire to know more. The explication of EQUAL is further expanded for any reader desiring more detail by looking to *Leviathan* chapters 1–11.

Having discovered the accident of all bodies in *this* system (EQUALITY) and their basic motions (PASSIONS), including the passion of HOPE, the experiment adds something back into to the scenario (step 3). We now consider the addition of competing appetites (and competing hopes) in a situation that has a relative scarcity of resources: "And therefore if any two men desire the same thing, which neverthelesse they cannot both enjoy, they become enemies" (Hobbes 2012, 190; 1651, 61). The result of such competing desires is that "diffidence" of one to another leads each individual to "secure himself by Anticipation" with the goal of being able to "master the persons of all men" (Hobbes 2012, 190; 1651, 61). This situation leads individuals to find "no pleasure [. . .] in keeping company" with one

112 HOBBES'S TWO SCIENCES

another and is "called Warre; and such warre, as is, of every many, against every man" (Hobbes 2012, 190–192; 1651, 61–62). The state of WAR also prevents the things necessary for commodious living, such as industry and navigation (Hobbes 2012, 192; 1651, 62). WAR thus is a feature of the privational scenario when we consider what results from the addition of competing desires and a relative scarcity of resources (not all desired things can be shared).[22]

Hobbes steps briefly outside of the thought experiment by considering a potential objection that the "inference" made in the thought experiment does not align with experience. Some individual, "not trusting this Inference, made from the Passions," might want to have it "confirmed by Experience" (Hobbes 2012, 194; 1651, 62). The "inference" that Hobbes has in mind is that by considering humans in their natural state (EQUALITY + HOPE + PASSIONS), those engaging in the thought experiment infer that this would be a state of war that is "every man, against every man" (2012, 190–192; 1651, 61–62). To those who do not trust that inference, Hobbes replies that it is confirmed by the fact that individuals lock their doors and chests. Given my claim that in a demonstration a thought experiment serves as a shortcut in place of an analysis, Hobbes's reply here might seem odd.

If all that is required to reach the same conclusion regarding humans' natural state is to have each person "consider within himselfe, when taking a journey, [that] he armes himselfe, and seeks to go well accompanied [. . .]" (Hobbes 2012, 194; 1651, 62), then why would Hobbes not simply have requested that the reader do that? I suggest that we understand Hobbes as amenable to reaching the same conclusion by multiple methods,[23] something he notes in *De corpore* VI.7:

> Civil philosophy is connected to moral [philosophy] in such a way that it can nevertheless be detached from it; for the causes of the motions of minds are not only known by reasoning but also by the experience of each and every person observing those motions proper to him only. (Hobbes 1981, 301; OL I.65; EW I.73)

Thus, one can reach the same conclusion—what Hobbes identifies as the "principles of civil philosophy," such as those that are being discussed in the present

---

[22] Hobbes also posits glory as one of the "three principall causes of quarrel" (2012, 192; 1651, 61). Given Hobbes's identification of glory with the "exultation of the mind" from "imagination of a mans own power and ability" and, if from considering past experiences, Hobbes claims that it is just "confidence" (2012, 88; 1651, 26–27), I suggest that Hobbes understands glory in the state of nature in terms of HOPE. One has the passion HOPE (i.e., the opinion of attaining what one desires) in the state of nature when one *also* has glory by recalling one's power and ability in past successes. Such glorying provides one with support for HOPE. For discussion of the importance of glory in Hobbes's political philosophy, see Slomp (2000).
[23] Hobbes is similarly sanguine to various methods in geometry being used for different purposes, even while preferring some over others: "But because there are many means by which the same thing may be generated, or the same problem constructed, therefore neither do all geometricians, nor doth the same geometrician always, use one and the same method" (EW I.312; see discussion in Jesseph 2017).

section—when one proceeds by synthesis from the starting points (starting with EQUALITY + HOPE and then adding PASSIONS) or when one who "has not learned the earlier part of philosophy, namely, geometry and physics" arrives at these principles by analysis from his or her everyday experiences (Hobbes 1981, 301; OL I.65; EW I.74). Such analysis according to the latter route might propose a question such as "whether such and such an act is just or unjust" (Hobbes 1981, 303; OL I.66; EW I.74). By resolution of 'unjust' into 'fact' and 'against the laws', and resolution of these further, Hobbes claims that "one finally arrives at the fact that the appetites of men and the motions of their minds are such that they will wage war against each other unless controlled by some power" (Hobbes 1981, 303; OL I.66; EW I.74).

Hobbes's reply to the individual who does not "trust" the "inference" in *Leviathan* 13 fits this latter route. Such an individual receives confirmation of the conclusion by checking his experience and seeing that the conclusion is grounded in claims about the appetites of human bodies. However, where does the state of nature thought experiment fall within the schema of *De corpore* VI.7? It seems neither a synthesis from the first principles of geometry or physics nor beginning from an analysis of everyday experience.

I suggest that we should understand the thought experiment as a third route to reach the same conclusion; the route one takes depends upon whether one needs to *discover, demonstrate*, or *receive confirmation*.[24] An individual who has "learned the earlier part of philosophy, namely, geometry and physics" (Hobbes would perhaps have himself in mind here, or Adam) could *discover* the principles of civil philosophy by means of an analysis to and synthesis from the first principles of civil philosophy (HUMAN devoid of CIVIL RELATIONS). When needing to *demonstrate*, that individual would omit the analysis and provide a thought experiment. The student (perhaps the reader of *Leviathan*) would gain an explication of the principles of civil philosophy by considering the experiment, but if needed they could analyze everyday experience and *receive confirmation* of the conclusion.

Although it has been suggested in chapter 3 (subsection 3.6.1) that Hobbes views the principles of civil philosophy as deriv*able* from the "earlier part of philosophy," i.e., geometry and physics, I noted that he nevertheless sees the principles of civil philosophy as simples in their own right. As I argued there, he concludes the first part of *Elements* (*Humane Nature*) immediately following the final article of chapter XIII with a summative claim that suggests this:

> Thus we have considered the nature of Man so far as was requisite for the finding out *the first and most simple elements* wherein the compositions of Politick Rules

---

[24] This notion of receiving confirmation adds to the two methods in *De corpore* VI discussed in chapter 3: the "method of invention" in *De corpore* VI.2–10 and the "method of teaching" in *De corpore* VI.11–19 (for discussion of these two methods, see Talaska 1988, 210).

114 HOBBES'S TWO SCIENCES

and Laws are *lastly resolved*; which was my present purpose." (1650, 170; emphasis added)[25]

Thus the conceptions that arise when considering human bodies in the state of nature, such as EQUALITY and HOPE, function as the simples for *the science of civil philosophy* and not as simples absolutely. Here Hobbes's commitment to the perspectival principle, discussed already in chapter 3, is evident. No particular "level," to use an anachronism, of external reality is explanatorily basic. When we are interested in seeing how human bodies arrive at peace, we may take human bodies outside of civil relationships as our starting point. This is what Hobbes means when he states humans in their natural state are the "first and most simple elements" of civil philosophy.

Returning to the experiment, *Leviathan* 13 concludes by claiming three additional things about human bodies. First, by determining the definition of 'injustice' we learn that in humans' natural state "nothing can be Unjust." This follows from consideration of the privational scenario because "Where there is no common Power, there is no Law; where no Law, no Injustice" (Hobbes 2012, 196; 1651, 63). Second, we learn that the way for human bodies to "come out of" this "ill condition" consists "partly in the Passions, partly in his Reason" (Hobbes 2012, 196; 1651, 63). Furthermore, the "Passions that encline men to Peace, are Feare of Death; Desire of such things as are necessary to commodious living; and a Hope by their Industry to obtain them" (Hobbes 2012, 196; 1651, 63). Considered as part of humanity's "natural condition," we have already understood the passions as the basic motions that move human bodies. When humans move from war to peace, then, we have learned that they do so by being moved by these three passions. A final property of human bodies in their natural state is that "every man has a Right to every thing; even to one anothers body" (Hobbes 2012, 198; 1651, 64). The right of nature is understood as "the Liberty each man hath, to use his own power [...] for the preservation of his own Nature" (Hobbes 2012, 198; 1651, 64).

In sum, in the state of nature thought experiment the reader considers human bodies apart from civil relations (step 1) and discovers that such bodies possess EQUALITY in the faculties mind and body and are moved by PASSIONS, including the passion of HOPE (step 2). Next, the reader learns that in such a state the addition of competing desires (step 3), including competing HOPE, leads to WAR, and that in this state each human body has LIBERTY. When such bodies are moved toward peace, they do so by being moved by the three passions mentioned above and by reason discovering the path to this end.

---

[25] Tönnies provides this "conclusion" in a footnote (Hobbes 1928, 191 fn. 11).

4.3 Making with Generative Definitions in Geometry and Civil Philosophy

In this section, I show that the simple conceptions explicated in the annihilation thought experiment and in the state of nature thought experiment are used in generative definitions of these two causal sciences. I suggest that in each of these *scientiae* an end is proposed and then the simples for that science are used to cause the end by means of generative definitions. Merely apprehending the simples and giving their explications would be insufficient; we must use them to make the ends we seek.[26]

4.3.1 "We ourselves draw the lines"

As mentioned already, Hobbes criticizes Euclid's definitions because they are not generative (EW VII.184). Generative definitions like "a line is made by the motion of a point" (Hobbes 1981, 297; OL I.63; EW I.70) "register" how to make a figure, but it is also necessary to consider the bodies used while making in a particular way. Hobbes's critique of Euclid is not entirely fair, since not all definitions in Hobbes's geometry include causes—some are explications. The first definition by explication in Hobbes's geometry is of 'point', understood as a body considered without magnitude. Generation enters into geometrical definitions with 'line', in which Hobbes uses explicated definitions for MOTION and POINT.[27] Further definitions build upon 'line': "a surface is made from the motion of a line," using MOTION again.

Hobbes views geometry as *scientia* because the geometer knows actual causes through the construction of figures. He connects this requirement for generative definitions to his understanding of demonstration as proceeding through syllogisms, arguing in *De corpore* VI.13 that the "goal of demonstration is the knowledge of causes and the generation of things; and if this knowledge is not in the definitions it cannot be in the conclusions of the syllogism [. . .]" (OL I.73; EW I.82–83). Thus, Hobbes's view that definitions must be causal follows from two claims: first, that conclusions of syllogistic demonstrations leading to *scientia* must themselves include causes; and second, that the only way for a conclusion to contain something is if it was already contained in the premises.

---

[26] See chapter 3, section 3.3, for discussion of this distinction between "apprehending" (*cognoscere*) and "putting together" to know scientifically (*scire*) (OL I.61; EW I.68).

[27] The relationship between the simple conception BODY and geometrical figures, such as 'line', may seem strange since it would seem that the conception BODY could be constructed from more basic geometrical notions. Thus it would seem that BODY could be analyzed further into lines, surfaces, and motion. However, Hobbes's desire to understand geometry as founded upon ways of considering bodies would prevent this inversion. This is why Hobbes denies that there is anywhere a breadthless line (EW VII.202) but only bodies that we consider as breadthless.

116   HOBBES'S TWO SCIENCES

Recall Hobbes's example of a SQUARE discussed in chapter 3. That example asks the reader to undergo an analysis from effects followed by a synthesis wherein one can "put [the parts] together as the cause of the square" (Hobbes 1981, 293; OL I.61; EW I.69),[28] but were Hobbes providing a *demonstration* of SQUARE to someone else, the analysis would be omitted and only the synthesis would be given (OL I.71; EW I.80). So a synthetic demonstration of how to make a SQUARE would begin from the simples of BODY and MOTION. Clearly, simply putting a point in motion would be insufficient, so four points must be moved to make lines in a certain way (straight lines) and in a certain proportion to one another (equality understood as equal magnitude). Then the four constructed lines must be placed together to make four right angles, which is done by making a circle and marking one-fourth of its perimeter.

How could a teacher's demonstration of how to "put together" the cause of SQUARE impart *scientia* to a student? As the student followed the teacher's instructions—the second "speciall use" of language (2012, 50; 1651, 13)—while making their own square (e.g., the teacher would say, "move that point to form a straight line"), they would gain causal knowledge of how their *particular* square was made, and insofar as they followed the teacher's directions regarding how to consider each part used in construction, they would gain *universal* knowledge of all squares (e.g., the teacher might say, "consider this body as a point," or "consider that body as a line without breadth"). The student would then "register" their own discovery of how to make SQUARE with a definition for 'square': "a square is made by moving four bodies, considered as points, to equal lengths and assembling those lines together to form four right angles and thus four sides." The student could remember this entire definition with the names 'equilateral', 'quadrilateral', and 'rectangular' or simply the single name 'square' since these are interchangeable according to Hobbes (Hobbes 1981, 321; OL I.75; EW I.85).

Would this demonstration of how to make a SQUARE be amenable to treatment with a syllogism as Hobbes requires for *scientia*? Hobbes does not provide any such syllogisms for constructions in geometry, but neither does he provide syllogisms in presenting the Laws of Nature. Nevertheless, could constructions of figures like the SQUARE be represented syllogistically? Such a demonstration would begin with definitions by explication of names like 'point' and 'motion'. After such definitions were provided, the following syllogism would capture the features of this demonstration:

1) All constructions beginning with four bodies, considered as points, moved in straight lines to equal lengths and assembled so those lines together form

---

[28] As a result, it is something like an analysis wherein one might ask "whether such and such an act is just or unjust" (Hobbes 1981, 303; OL I.66; EW I.74).

DEMONSTRATING SCIENTIFIC KNOWLEDGE   117

four right angles and thus four sides are constructions of figures that are equilateral, quadrilateral, and rectangular.

2) All constructions of figures that are equilateral, quadrilateral, and rectangular are constructions of squares.

Ergo: All constructions with four bodies, considered as points, moved in straight lines to equal lengths and assembled so those lines together form four right angles and thus four sides are constructions of squares.

In accordance with Hobbes's understanding of a cause as an "entire cause [*causa integra*]" (OL I.107–108; EW I.121–122), the syllogism above provides the necessary and sufficient conditions to make a SQUARE, and, in doing so, it represents what is in the mind "when a person remembers that those attributes are all one and the same thing, he concludes that those three names are also names of the same thing; that is that the conclusion is true" (Hobbes 1981, 259; OL I.44; EW I.49).[29] Since Hobbesian names can be a single vocable like 'square', or more than one like 'equilateral', 'quadrilateral', and 'rectangular' (Hobbes 1981, 205; OL I.17; EW I.19), anyone considering the conclusion recognizes that a SQUARE has all of the attributes described and thus that it is true.

4.3.2 "We ourselves make the principles—that is, the causes of justice (namely laws and covenants)"

At the close of *Leviathan* 13, Hobbes asserts that both reason and the passions cause human bodies to move from their natural state to peace. The three passions responsible for this motion are "Feare of Death; Desire of such things as are necessary to commodious living; and a Hope by their Industry to obtain them" (Hobbes 2012, 196; 1651, 63). However, these passions, as endeavors, say nothing about *how* to reach the end toward which their motion is directed. If just these passions were provided in the synthetic demonstration, we would be left in a situation similar to being instructed as follows: "make a square by moving a point to make some number of lines." Like the causes of a square in addition to 'line'—four straight lines of equal length linked by right angles—the Laws of Nature *direct* the motion of the three passions. Thus reason "suggesteth convenient Articles of Peace [. . .] which otherwise are called the Lawes of Nature" (Hobbes 2012, 196; 1651, 63). The Laws

---

[29] In contrast, representing an analysis would require a "series of syllogisms" (cf. Hobbes 1981, 323; OL I.76; EW I.86) each of which would specify a necessary condition. For example, *one* syllogism in such a series could be as follows: Premise 1: All squares are rectangular figures; Premise 2: All constructions of rectangular figures are constructions with four right angles; Ergo: All squares are constructions with four right angles. Individually, none of these syllogisms in the series would be sufficient to construct a square but each would be necessary. In synthesis, one puts all of these necessary conditions together (adding/compounding) and thus shows how to make the thing in question.

of Nature, in other words, direct the motion so that we can satisfy the end given by these three passions. In this section, I suggest that the Laws of Nature, considered jointly, play the same role as the components of SQUARE that we "put together" to cause it.

I shall assume that the three aforementioned passions provide the end toward which human bodies in their natural state are moved—peace. As Hobbes says, these passions "encline men to Peace" (Hobbes 2012, 196; 1651, 63). This is analogous to having an end in mind when working in geometry. As a demonstration, we should expect the presentation of the Laws of Nature to lack an analysis. Thus, we should anticipate that Hobbes would provide only a synthesis from explicated simples and, like in geometry, we should understand this "putting together" in synthesis as showing the cause of peace. What are the Laws of Nature on this view? In short, they "specify an optimum set of actions designed to bring about peace" (Malcolm 2002, 32). Given that you desire peace, do what the Laws of Nature specify.[30] Likewise, given that you desire to construct a square (and not, say, a rhombus), follow the instructions provided by the "cause" of SQUARE.

To provide evidence for my suggestion that the Laws of Nature play the same role as the component parts out of which we cause a square, I shall consider several of the Laws of Nature. To recall the earlier discussion from subsection 4.2.2, the basic properties of human bodies in their natural state are that they possess EQUALITY, they are moved by PASSIONS, and that EQUALITY gives rise to the passion of HOPE. We learn in *Leviathan* 14 that all human bodies in their natural state have LIBERTY. Finally, WAR arises because of conflicting PASSIONS, including HOPE.

The first law is introduced by appealing to LIBERTY, and it uses the simples of HOPE and WAR—motions of human bodies (HOPE) and the conflict arising from those motions frustrating each other (WAR)—that were explicated in the thought experiment:

> [...] as long as this naturall Right of every man to every thing endureth, there can be no security to any man, (how strong or wise soever he be), of living out the time, which Nature hath ordinarily alloweth men to live. And consequently it is a precept, or generall rule of Reason, *That every man, ought to endeavour Peace, as farre as he has hope of obtaining it; and when he cannot obtain it, that he may seek, and use, all helps, and advantages of Warre.*" (Hobbes 2012, 198–200; 1651, 64)

The second law, which is "derived" from the first law, also uses LIBERTY; it instructs that insofar as human bodies seek peace, they must relinquish the LIBERTY found

---

[30] Hoekstra (2003, 116) similarly sees the laws of nature as conditional and argues Hobbes views the necessary conditions as satisfied for all humans.

in the state of nature and "lay down this right to all things" (Hobbes 2012, 200; 1651, 64–65).

Following this presentation of the first two laws, Hobbes clarifies what laying down a right entails, the nature of injustice, and the distinction between contracts and covenants. All of these distinctions occur to show how the machinery required for the first and second law is present in human bodies' natural state, and this is the reason that Hobbes places them in Part I of *Leviathan* ("Of Man"). This concords with the Table in *Leviathan* IX, where, for example, "contracting", which delivers "The *Science of the Just and Unjust*," is a consequence of speech, which itself is one of the "Consequences of from the Qualities of *Men in speciall*" (Hobbes 2012, 130–131). Instances of contracting are consequences that follow from humans considered as bodies.

The simples explicated in the state of nature thought experiment are used in various other Laws of Nature. In the discussion of the third law, "*That men perform their covenants made*" (Hobbes 2012, 220; 1651, 71), we find that without this instruction human bodies will return to the condition of WAR. In the ninth and tenth laws, Hobbes directly appeals to the simple of EQUALITY. He articulates the ninth law as against pride: "*That every man acknowledge other for his Equall by Nature. The breach of this Precept is Pride*" (Hobbes 2012, 234; 1651, 77). Hobbes asserts a relationship between the ninth and tenth laws, saying that upon the ninth "dependeth another." The ninth law is "*That at the entrance into conditions of Peace, no man require to reserve to himselfe any Right, which he is not content should be reserved to every one of the rest*" (Hobbes 2012, 234; 1651, 77). Here again he appeals to the simple of EQUALITY: "[. . .] the acknowledgement of naturall equalitie" (Hobbes 2012, 234–236; 1651, 77). EQUALITY recurs in the eleventh law, where judges are instructed to "deale Equally", in the twelfth law, where things that cannot be divided must be "enjoyed in Common", and in the thirteenth law, where for the sake of "equall distribution" lot is the means by which things which cannot be divided or enjoyed in common are to be distributed (Hobbes 2012, 236; 1651, 78).

Importantly, the Laws of Nature are generative definitions insofar as they tell human bodies what to do to cause peace. As we have seen earlier in *De homine* 10.5: "We ourselves make the principles—that is, the causes of justice (namely laws and covenants) [. . .]" (OL II.94; Hobbes 1994b, 42). In this way, they are analogous to the causes of SQUARE. If one wants to make PEACE, one should do what is specified by the Laws of Nature, considered jointly. Furthermore, just like the individual who has never seen a circle made may doubt whether "such a figure [is] possible" (EW VII.205), prior to seeing all of these parts (the Laws of Nature) put together, the student may have doubted that it was possible to make PEACE (recall the "argument from the Practise of men, that have not sifted to the bottom" discussed in section 3.2 of chapter 3; cf. 2012, 320; 1651, 107). Just like determining the properties "from the known generation of the displayed figure, [is] most easy" in the case of a circle, but not possible by "sense" (Hobbes 1981, 181; OL I.5;

120   HOBBES'S TWO SCIENCES

EW I.6), the student who has constructed PEACE by putting together all of its parts knows all of its properties.

Thus, as in the case of the square, once we "put together" all of the Laws of Nature, we know the cause of PEACE. Indeed, Hobbes indicates that he considers this as adding together (compounding/synthesis) when he gives the single rule that will enable all humans to "examine" the Laws of Nature. Hobbes holds that all of the laws can be "contracted into one easie sum [. . .] *Do not do to another, which thou wouldest not have done to thy selfe*" (Hobbes 2012, 240; 1651, 79). In addition to this language of "sum", which suggests that we are to put together by reckoning all of the Laws to gain peace, Hobbes uses the metaphor of the "ballance":

> [. . .] he has nothing more to do in learning the Lawes of Nature, but, when weighing the actions of other men with his own, they seem too heavy, to put them into the other part of the ballance, and his own into their place, that his own passions, and selfe-love, may adde nothing to the weight; and then there is none of these Lawes of Nature that will not appear to him very reasonable. (Hobbes 2012, 240; 1651, 79)

This reference to the balance, and the simple of EQUALITY implicit in it, might have seemed figurative had not EQUALITY played such a central role in Hobbes's preceding discussion of the laws.

Would this demonstration of how to make PEACE be amenable to treatment with a syllogism as Hobbes requires of a demonstration? After providing definitions by explication for the simple conceptions (EQUALITY, PASSIONS, HOPE, LIBERTY, and WAR), Hobbes would likewise provide a definition for PEACE, which he does within the thought experiment by showing its relationship to WAR: "[. . .] the nature of War, consisteth not in actuall fighting; but in the known disposition thereto, during all the time there is no assurance to the contrary. All other time is PEACE" (2012, 192; 1651, 62). Next Hobbes must make explicit that all human bodies in their natural state have passions that "encline" toward the end of PEACE (2012, 196; 1651, 63). Given this initial set up, Hobbes would offer the following syllogism to represent the synthetic demonstration of the laws of nature:

1) All situations in which [all of the Laws of Nature are performed] are situations absent of war.
2) All situations absent of war are situations of peace.
   Ergo: All situations in which [all of the Laws of Nature are performed] are situations of peace.

This brief syllogism represents the approach Hobbes *actually* took in presenting a synthetic demonstration of all of the Laws of Nature in *Leviathan* 14–15, which he held could be compacted into "one easie sum." Imagine if Hobbes's reader (the student of the demonstration he offers) were to object: "What if only the first, second, and third laws were performed? Would that not be sufficient to generate peace?" Hobbes's reply would be that though the first three Laws are *necessary* for PEACE they are not sufficient to generate it since, for example, in such a situation individuals would be at liberty to violate laws like the ninth law and reserve certain rights for themselves while withholding them from others (Hobbes 2012, 234; 1651, 77).[31] Were the student asked if the conclusion of the syllogism is *true*, they would need to determine whether war would be possible were all the Laws of Nature performed. As in the case of any Hobbesian syllogism with a true conclusion (Hobbes 1981, 259; OL I.44; EW I.49–50), remembering that the Laws of Nature considered together name the same thing as 'peace', i.e., absence of 'war', would show them that the conclusion is true.

4.4 Benefits of this Account of Hobbesian Demonstration

I have argued that the Laws of Nature in civil philosophy play the same role as generative definitions of geometrical figures, such as the definition of 'square'. The link between civil philosophy and geometry is natural given Hobbes's pronouncement that they are both *scientiae*—two sciences in which we possess actual causal knowledge. All other knowledge is from the effects to possible causes. However, even though Hobbes requires generation to be in the definitions, not all definitions can be causal. Some definitions are explications of simple conceptions.

This chapter has furthermore argued that for Hobbes the definitions by explication for these simple conceptions are achieved in a thought experiment. The two thought experiments considered—the annihilation thought experiment and the state of nature thought experiment—both aim at providing definitions by explication and then these explicatory definitions are used within the generative definitions. The structure in civil philosophy and geometry of a thought experiment, definitions by explication, and generative definitions allows Hobbes to see himself as providing a demonstration by synthesis in both cases. The synthesis begins with first principles, the simples, and ends by putting together all the component parts to make the thing in question, whether SQUARE or PEACE. Each

---

[31] Like in the case of SQUARE (see fn. 29 above), if Hobbes were representing an analysis from a peaceful commonwealth to simple conceptions, he would represent each of the necessary conditions for PEACE with a single syllogism and use a series of such syllogisms. For example, Premise 1: All situations of peace are situations absent of war; Premise 2: All situations absent of war are situations in which the Third Law of Nature is performed; Ergo: All situations of peace are situations in which the Third Law of Nature is performed. As in the example of SQUARE, Hobbes's goal in *Leviathan* is to show how to make PEACE and thus he would represent that demonstration syllogistically by including the necessary *and* sufficient conditions to generate it.

122 HOBBES'S TWO SCIENCES

synthetic demonstration is amenable to representation with a syllogism, but Hobbes's own manner of demonstration does not follow this strictly in practice. Instead, he simply provides all of the individually necessary, jointly sufficient conditions for constructing the ends in question, and the reader is to add them all together in their mind. According to Hobbes, this very activity is "what in the mind might answer to a syllogism" (Hobbes 1981, 259; OL I.44; EW I.49).

In this section, I describe two advantages of the understanding of the laws of nature for which I have argued and contrast the maker's knowledge view with three influential views of the Laws of Nature: the Laws of Nature as a demonstration from an axiom, the Laws of Nature as God's commands, and the reciprocity interpretation of the Laws of Nature. First, understanding Hobbesian civil philosophy as Making Type 3 attends to Hobbes's geometrical practice rather than assuming his geometry, and as a result his civil philosophy, is axiomatic in nature. Rather than presenting the reader with axioms, Hobbes offers thought experiments to provide definitions by explication that will serve as the first principles of a synthetic demonstration. As mentioned in in chapter 1, Hobbes's supposed affection for Euclid is as old as Aubrey's description of Hobbes's accidental encounter with an open copy of the *Elements* in a library that "made him in love with geometry" (Aubrey 1898, 332). However, seeing Aubrey's report as providing evidence of Hobbes's fondness for Euclid is not without difficulties, which becomes evident when examining Hobbes's views of Euclid, as well as Hobbes's own practice of geometry. Indeed, Hobbes criticizes some of Euclid's definitions since they are not generative.

At other points in the corpus, Hobbes is critical of features of the Euclidean program where axioms are assumed without demonstration: "For the axioms which we have from Euclid, which are possible to demonstrate, are not principles of demonstration [. . .]" (*De corpore* VI.13; OL I.72; EW I.82). Rather than assuming axioms, perhaps on the authority of the teacher or on account of their self-evidence, Hobbes holds that even those axioms must be demonstrated: "[. . .] to the end that the reader may know that those axioms are not indemonstrable, and therefore not principles of demonstration; and from hence learn to be wary how he admits any thing for a principle, which is not at least as evident as these are" (*De corpore* VIII.25, OL I.105–106; EW I.119). Instead of beginning from axioms, this chapter has endeavored to show Hobbes grounding synthetic demonstrations, whether in geometry or civil philosophy, in definitions of simple conceptions explicated in a thought experiment. These explications are then used in generative definitions that give the cause of our constructions. Such considerations weigh against Deigh's "definitivist" understanding of the laws of nature (1996; 2003) wherein one begins with the definition of a Law of Nature in *Leviathan* 14 as an axiom, and from that axiom derives the Laws of Nature found in chapters 14 and 15. The maker's knowledge understanding of the laws of nature shows how Hobbesian civil philosophy is structured according to Hobbes's own *more geometrico* without committing it to an axiomatic foundation.

A second benefit to seeing Hobbesian civil philosophy as Making Type 3 is that it can straightforwardly make sense of Hobbes's statement in *Leviathan* 15 that the Laws of Nature are "Immutable" and "Eternal":

> The Lawes of Nature are Immutable and Eternall; For Injustice, Ingratitude, Arrogance, Pride, Iniquity, Acception of persons, and the rest, can never be made lawefull. For it can never be that Warre shall preserve life, and Peace destroy it. (Hobbes 2012, 240; 1651, 79)

In seeing the Laws as playing the same role as the cause of SQUARE, we can make sense of this claim without requiring some lawgiver, such as God, to vouch for their immutability or eternality. In just the same way that "it can never be that Warre shall preserve life," it can never be the case that a square is composed of curved lines or unequal sides. We might have imposed different names, perhaps signifying the conception SQUARE with the name 'donkey', but insofar as the conception SQUARE is considered, Hobbes can coherently hold that it could never be composed otherwise than as the definition that he provides. And indeed, Hobbes himself denies that the Laws of Nature are properly laws since "Law, properly, is the word of him, that by right hath command over others" (2012, 242; 1651, 80). Instead, Hobbes holds that these "laws" are "but Conclusions or Theoremes concerning what conduceth to the conservation and defence" and as such they are "dictates of Reason" (2012, 242; 1651, 80). In the end, the Laws of Nature are like the causes of SQUARE: they tell us how to make the peace that we desire.

Insofar as the Laws of Nature play the role of the parts of Hobbesian geometric definitions, the maker's knowledge view thus holds that the laws exist only insofar as humans exist and their desires are aimed at PEACE. Even though Hobbes held that all humans in fact desire PEACE (agreeing with Hoekstra 2003), the Laws exist only insofar as that contingent fact holds. In other words, despite Hobbes's use of language like "immutable" to describe them, they do exist apart from God's command. Indeed, he says that we can describe them as "Conclusions, or Theoremes" but that we may also call them "Lawes" if we "*consider* the same Theoremes, *as* delivered in the word of God" (2012, 242; 1651, 80; emphasis added). This language of "consider as" is directly explainable on the maker's knowledge view, since on my account Hobbes's epistemology itself is founded in this very cognitive ability. In short, the maker's knowledge view can accommodate the possibility of *considering* the Laws of Nature *as* "delivered in the word of God," but it does not hold that this *must* be the case and thus avoids the problems faced by viewing the laws as God's commands.[32]

---

[32] Gauthier (2001, 262) suggests that the laws of nature are first dictates of reason and then later come to be known either by being enshrined in Scripture or as part of civil laws.

124 HOBBES'S TWO SCIENCES

Indeed, viewing the Laws of Nature as God's commands is a feature of Martinich's account in *The Two Gods of Leviathan* (2002). A worry faced by Martinich's view is that the Laws of Nature are not in the form of commands—their logical form is not as imperatives. Martinich recognizes the propositional form of the laws, rather than the imperative, and takes this to be explained by understanding the Laws as "derivations," and he contrasts them with "geometrical propositions that are derivable from reason but do not have any obligatory character" (Martinich 2002, 121–122). Taking the Laws of Nature as "derived," in a way similar to Lloyd's view (discussed below), leads Martinich to misrepresent the character of Hobbesian demonstration. The maker's knowledge view does not require holding that they are commands but recognizes that the Laws are simply instructions to perform some action, just like the generative geometrical definitions that Hobbes provides. They are no different in logical form than what one finds in instruction manuals—given that one desires some end, the following actions should be performed. Thus, a benefit of the maker's knowledge view over this account is that it not only shows how Hobbes could view civil philosophy as *scientia* (like geometry) but avoids forcing the Laws into a logical form other than the one in which Hobbes provided them.

One of the aims of Lloyd's reciprocity interpretation is to show how the Laws are normatively binding for all rational agents. Lloyd's interpretation is novel and compelling, but it fails to appreciate Hobbes's own views of demonstrations as synthetic constructions because it does not engage with Hobbes's distinction in *De corpore* between the "method of discovery" (discussed in chapter 3, sections 3.2–3.5) and the "method of teaching" (discussed above). In *Morality in the Philosophy of Thomas Hobbes* (2009), and in later works, Lloyd argues that the Laws are derived from the concept "accordance with reason," which she identifies as a theorem concerning reciprocity. The reciprocity theorem asserts that one acts contrary to reason if one judges another's act as being done without right but then does that action oneself, and the Laws are derived from this theorem with the help of premises that are analytic or confirmed by what Lloyd calls "indubitable introspectables." To buttress this account, Lloyd asserts that Hobbes viewed demonstrations "by definition, [as] the method of establishing conclusions by syllogistic inference from definitions" (Lloyd 2009, 224) and gestures to chapters I–VI of *De corpore* as support for this significant, and I argue mistaken, assumption regarding Hobbes's method of demonstration. Rather than characterizing them as a "derivation," the maker's knowledge account of the Laws understands the presentation of the laws as a synthetic demonstration beginning from simple conceptions. This account both mirrors his annihilation thought experiment in *De corpore* and is consistent with Hobbes's explicit discussions of the nature of demonstration.

In sum, in contrast to these three others—the definitivist, Laws as God's commands, and reciprocity views—the maker's knowledge view maps onto Hobbes's explicit comments about the nature of demonstration and method while simultaneously permitting syllogistic representation of the role of the Laws of

Nature in causing PEACE. A synthetic demonstration provides a maker with the individually necessary, jointly sufficient conditions for constructing some end, so that she can add them all together in her mind. This very activity is "what in the mind might answer to a syllogism" (Hobbes 1981, 259; OL I.44; EW I.49–50). At the same time, the maker's knowledge view does this neither by seeing derivation as Hobbes's method of demonstration nor by forcing the Laws of Nature individually into the logical form of imperatives. Representing Hobbes's Laws of Nature syllogistically is an aim of all three views mentioned above, but each makes problematic assumptions beyond Hobbes's own views to do it. While in actual practice Hobbes does not provide syllogisms for constructions in geometry or for the Laws of Nature in civil philosophy (something that he would likely have done if he thought it so important), the maker's knowledge view can nevertheless accommodate representing synthetic demonstrations as syllogisms (as I do in subsections 4.3.1 and 4.3.2).

4.5 The Relationship between *Leviathan* Parts I and II

It is now possible to reevaluate the relationship between the first two parts of *Leviathan*—Part I "Of Man" and Part II "Of Commonwealth." In chapter 1, I raised a worry regarding this relationship for the deductivist understanding of Hobbes's philosophy. Since Hobbes says in *De corpore* VI.7 "[c]ivil is connected to moral [philosophy] in such a way that it can nevertheless be detached from it" (Hobbes 1981, 301; OL I.65; EW I.73), he seems to allow for individuals simply to study the motions of their own mind and gain knowledge of the principles of civil philosophy. Such a detachment is difficult to explain for the deductivist view because it makes it seem that Hobbes thought that one could develop civil philosophy entirely by introspection and apart from moral philosophy, natural philosophy, and first philosophy.

Nevertheless, *Leviathan* Part I *seems* to rely upon natural-philosophical claims insofar as it concerns sensation, imagination, memory, and the passions, among other topics. However, the analogy between Hobbes's two sciences, which is captured by the maker's knowledge view, aids in thinking about Hobbes's aim in *Leviathan* Part I in contrast to Part II. Just like mathematical objects are founded in simplest conceptions gained from sense, so also the simples of civil philosophy are *from* sense but are not claims in natural philosophy as such (natural philosophy is the subject of chapter 5). In other words, Part I is not about human bodies in the natural world as such, but about conceptions of human bodies considered as in special ways, akin to how we consider natural bodies in special ways for the objects of geometry. A consequence of this understanding of Parts I and II of *Leviathan* is that Hobbes would view *De corpore* Part IV as his primary contribution to natural philosophy, not *Leviathan* Part I.

126   HOBBES'S TWO SCIENCES

Leviathan Part I is thus Hobbes's construction from simples of PEACE, which culminates in the opening chapter of Part II, *Leviathan* 17, with the generation of the COMMONWEALTH and the definition of 'commonwealth'. Importantly, *Leviathan* 16 on persons and authors, which follows the two chapters on the Laws of Nature, is in Part I (likewise, *Leviathan* 12 on religion is in Part I, but I wait to discuss this until chapter 6). Although scholars have often treated "authorization" as a topic treated within Hobbes's civil philosophy,[33] he did not introduce it within the discussion of COMMONWEALTH. Personation, whether natural or artificial, and the distinction between an actor and an author, are ways of moving for human bodies in their natural state; these are, as it were, the motions of human bodies that allow for the possibility of forming the commonwealth later in *Leviathan* 17 and thus must be in place within Part I before Part II can begin. The generation of COMMONWEALTH thus combines all the features of Part I—humans considered as in their natural state, the construction of PEACE from simples through to the Laws of Nature, and the authorization procedure to give up one's rights—in the well-known conditional covenant: "*I Authorise and give up my Right of Governing my selfe, to this Man, or to this Assembly of men, on this condition, that thou give up thy Right to him, and Authorise all his actions in like manner*" (2012, 260; 1651, 87). Hobbes is clear that this generation provides the "only way" to secure PEACE, which follows directly from understanding the Laws of Nature as a synthetic demonstration showing the individually necessary, jointly sufficient conditions.

Hobbes next offers a definition of 'commonwealth' that signifies the essence of COMMONWEALTH: "*One Person, of whose Acts a great Multitude, by mutuall Covenants one with another, have made themselves every one the Author, to the end he may use the strength and means of them all, as he shall think expedient, for their Peace and Common Defence*" (2012, 260–262; 1651, 88). Through a series of transformations, Hobbes has led the reader from the synthetic demonstration that the "only way" to make PEACE is to perform all of the Laws of Nature, but now he adds that *continued* presence of PEACE after its construction is possible only if the group reduces "all their Wills, by plurality of voices, unto one Will" (2012, 260; 1651, 87). This unity in a single will provides "something more than Consent, or Concord" but instead a "reall Unitie of them all" (2012, 260; 1651, 87).

Leviathan Part II thus begins by finalizing the causal account of making PEACE, but the chapters that follow *Leviathan* 17 do something beyond this. I suggest that after providing the definition of 'commonwealth' Hobbes demonstrates properties of COMMONWEALTH that are consequences of its essence, such as the rights of sovereigns that he describes in *Leviathan* 18 and the liberty of subjects in *Leviathan* 21. Indeed, Hobbes reflects upon the structure of Part II by noting that it first treats

---

[33] As indicated, for example, by the location of a chapter on authorization within the "political philosophy" section in the Table of Contents of the *Oxford Handbook of Hobbes* (2016).

the "Generation, Forme, and Power of a Commonwealth" and then moves to considering its "parts," which he compares to parts of natural bodies (2012, 348; 1651, 115). This is like showing someone how to make a circle, providing a generative definition, and next focusing on a part of CIRCLE such as DIAMETER or RADII. The chapters of Part II following chapter 17 are thus analogous to corollaries in geometry insofar as something has already been made (COMMONWEALTH or TRIANGLE) in Part I and then in Part II something else is shown to follow from its essence.

In Hobbes's geometry, after one makes a figure and defines its essence, then one can demonstrate additional features that follow from that essence but are not contained within it. Hobbes's suggestion that "[f]rom this Institution of a Common-wealth are derived all the *Rights*, and *Facultyes* of him, or them, on whom the Soveraigne Power is conferred by the consent of the People assembled" (2012, 264; 1651, 88) supports this understanding of Part II. It follows in this way from the definition of 'commonwealth' that a reader sees in *Leviathan* 19 that the present sovereign holds the right to determine a successor (2012, 300; 1651, 100) and in *Leviathan* 20 that the power of the sovereign must be absolute, and thus is not "too great," for otherwise it would result in the sovereign having to submit to a greater power (2012, 320; 1651, 107). Another example includes the account in *Leviathan* 21 of the "Liberty to disobey" any command of the sovereign to harm oneself, fail to defend oneself, or starve oneself, which follows, as Hobbes says, from the necessity that any contract that requires such actions is void (2012, 336; 1651, 111). This invocation of the earlier demonstration in *Leviathan* 14 (2012, 202, 1651, 65–66) concerned the nature of what rights could be transferred and what could not, but Hobbes's explicit citation of it here is not simply to repeat what has been said already. The connection between that earlier demonstration and what I am suggesting is like a geometric corollary is as follows: while the *Leviathan* 14 shows that contracts are void under such conditions, which is part of the construction of PEACE that leads to COMMONWEALTH, and thus is contained within the essence of 'commonwealth', the claim Hobbes makes in *Leviathan* 21 about 'liberty' is not contained in that essence, nor in the claim about a particular void status. Nevertheless, the claim about liberty follows necessarily from the essence 'commonwealth' since there is "absence of opposition" to it in that definition. In short, since one could not have covenanted to transfer this right, there can be no opposition to exercising it. Insofar as there is no opposition to exercising it within COMMONWEALTH, it follows that there is a liberty.

Likewise, in Hobbesian geometry, one can demonstrate corollaries that follow from the nature of some figure once one has constructed it. For example, Hobbes does this when he demonstrates that "the three angles of a straight-lined plain triangle are equal to two right angles" in *De corpore* XIV.12, Corollary IV (OL I.165; EW I.109) (I discuss this corollary at length in chapter 3). Similarly, after defining 'simple circular motion' in *De corpore* XXI.1 with a construction to show that all

128   HOBBES'S TWO SCIENCES

straight lines in a body moving with that type of motion will always be kept parallel to themselves, Hobbes provides two corollaries that follow from that account of the essence of simple circular motion (OL I.258–260; EW I.317–319). In the same way, I suggest that the rights of the sovereign, the liberties of subjects, and other topics in Part II are meant to follow as corollaries from the nature of COMMONWEALTH.

Furthermore, the view that these chapters in Part II of *Leviathan* should be understood as akin to corollaries in geometry makes sense of why in these chapters Hobbes rejects any criticisms of his account from experience (i.e., the so-called objection "of the Practise" in *Leviathan* 20; 2012, 320; 1651, 107). Any appeal to the particularities of experience as a supposed objection to claims in Part II would fail to see the necessity that Hobbes claims is present in the account of COMMONWEALTH that he offers. Given the foundation in maker's knowledge, talk of SQUARE, PEACE, COMMONWEALTH, and TRIANGLE is unassailable by any appeal to experience. Beyond the "making" of PEACE, and the transformations that follow it to see COMMONWEALTH as what provides and maintains PEACE, this connection between corollaries in geometry and *Leviathan* Part II extends what it means to make in civil philosophy. Constructing in the corollaries in geometry includes making something new, such as a new line that is parallel to the base of a triangle, as discussed in chapter 3, section 3.5, and similarly so does considering what liberties follow for subjects in *Leviathan* Part II.

4.6 Conclusion

This chapter has argued that Hobbes understood his two sciences—geometry and civil philosophy—as instances of maker's knowledge. I have argued that these two sciences are instances of what I identify in chapter 3 as Hobbesian Making Type 3, which occurs when in teaching (*docere*) one skips a resolution, offers a thought experiment, and in synthesis constructs from that foundation. Given this status, makers in these two sciences have access to the actual causes that bring about the productions they create. They know, for instance, not only *that* all situations absent of war are situations of peace but more importantly *how* to create such situations because they possess causal (*why*) knowledge. When demonstrating to others, the geometer and civil philosopher both begin from a foundation where the starting principles—the simples for a given domain—are explicated in a thought experiment. Teachers proceed from those simples and make a geometrical figure or peace. Such demonstrations involve leading a student through the steps of construction and, in doing so, adding together all of the necessary components to generate the effect. Language is used throughout the demonstration to show students how to consider as and also to help remember the steps involved; syllogisms can be used to represent the component parts that are put together.

5

Hobbesian Natural Philosophy as Mixed Mathematics

At several points in the corpus, Hobbes claims that he has provided a unified system, with connections between geometry and natural philosophy.[1] As described in chapter 1, Hobbes's descriptions of this system have led some scholars to see this unity resulting from deductive connections between geometry and natural philosophy.[2] However, examining Hobbes's actual practice in natural philosophy belies taking his descriptions as conclusive. In this chapter, I offer an alternative account of the relationship of Hobbesian geometry to natural philosophy by arguing that mixed mathematics provided Hobbes with a model for thinking about it. In mixed mathematics, one may borrow causal principles from one science and use them in another science without a deductive relationship. Natural philosophy for Hobbes is mixed because an explanation may combine observations from experience or experiments with causal principles from geometry. As he puts it in *De homine* X.5, "[...] physics (I mean true physics), that depends on geometry, is usually numbered among the mixed mathematics" (Hobbes 1994b, 42). In the *practice* of Hobbesian natural philosophy, one may appeal to everyday experience or experiments for the demonstration that something happens (the 'that') and borrow the cause (the 'why') from geometry.

My argument shows that Hobbesian mixed natural philosophy is constrained by two *a priori* principles of motion from *De corpore* VIII.19 and IX.17 and then relies upon suppositions that bodies plausibly behave according to borrowed causal principles from geometry, acknowledging that bodies in the world may not behave this way. For example, Hobbes develops an account of "simple circular motion" in geometry and supposes that the sun moves the air around it by this type of motion. The natural philosopher does not know as a matter of fact that the sun causes this sort of motion but *supposes* that it does—Hobbes describes this as a "possible cause"—and then explains various phenomena related to light

---

[1] For example, in *De corpore* VI.6 Hobbes links what he calls "our simplest conceptions," such as 'place' and 'motion', with generative definitions in geometry and, ultimately, with natural philosophy and morality (OL I.62–65; EW I.70–73).

[2] Chapter 1 distinguishes between the "deductivist" and the "disunity" interpretations of Hobbes's philosophy. Advocates of the deductivist view include Hampton (1986), Martinich (1999; 2005), Peters (1956), Shapin and Schaffer (1985), and Watkins (1973). Others have seen Hobbes's philosophy as disunified (Robertson 1886; Taylor 1938; Warrender 1957; for discussion, see also Sorell 1986, 6).

Hobbes's Two Sciences. Marcus P. Adams, Oxford University Press. © Marcus P. Adams 2025.
DOI: 10.1093/9780198924715.003.0005

130   HOBBES'S TWO SCIENCES

and heat using it. As part of geometry, principles about simple circular motion have certainty; the philosopher can know that simple circular motion has necessary effects. However, when borrowing causal principles related to simple circular motion within a natural-philosophical explanation, one cannot know whether the sun actually operates by simple circular motion. As a result, natural philosophical explanations are suppositional: *if* the sun causes simple circular motion, *then* an effect of that propagated motion will be heat and light.

My argument proceeds in two stages. First, I consider Hobbes's comments concerning and relation to Aristotelian mixed mathematics and to Isaac Barrow's later broadening of mixed mathematics in *Mathematical Lectures* (1685), contrasting these affinities to Hobbes's contemporary John Wallis's understanding. I show that, for Hobbes, maker's knowledge from geometry provides the 'why' in mixed-mathematical explanations, and any explanation must be constrained by the two *a priori* principles of motion in *De corpore* VIII and IX.[3] Next, I examine three Hobbesian explanations: 1) the explanation of sense in *De corpore* XXV.1–2; 2) the explanation of the swelling of parts of the body when they become warm in *De corpore* XXVII.3; and 3) the explanation of two behaviors of the air-pump in *Dialogus Physicus, sive De natura Aeris* (1661; hereafter *Dialogus Physicus*). In these three explanatory contexts, I show Hobbes borrowing and citing geometrical principles and mixing those principles with appeals to experience.[4] The chapter

---

[3] Hobbes uses "mixed mathematics" (*mathematicas mixtas*) in *De homine* X.5 and in *Anti White*. For general discussion of "mixed mathematics", see Brown (1991). I have discussed making and *scientia* in chapter 3. My aim in linking Hobbesian natural philosophy with mixed mathematics is not to locate the genesis for this feature of Hobbes's thought, nor is it to delineate Hobbesian mixed-mathematical natural philosophical explanations from the myriad predecessors to whom Hobbes may have looked, such as Kepler, Copernicus, Galileo, and others. Like many other Early Modern philosophers, Hobbes often did not reveal his sources. For discussion of some of these predecessors, see Dear (1988) and Machamer (1978). Sections 5.1 and 5.2 are drawn from parts of Adams (2016a), and section 5.3 is drawn from parts of Adams (2017). Biener (2016) also uses mixed mathematics as a way of understanding the structure of Hobbes's philosophy. Raylor (2018, 210–212) contrasts Hobbes's comments about natural philosophy as conjectural with his view that geometry and civil philosophy are demonstrable because humans construct the objects in those disciplines. Raylor points to a tension: while natural philosophy is characterized by its "non-demonstrability [and] conjecturalism," like the Scholastic discipline of meteorology, Hobbes nevertheless places it within the table of the sciences in *Leviathan* IX (Raylor 2018, 211). Raylor suggests that this tension can be assuaged by Hobbes's identification of natural philosophy, and physics broadly, as knowledge of the *appearances* that, as such, leads only to "some knowledge" and not, unlike geometry and civil philosophy, knowledge of *actual* causes. However, Raylor's account only explains why natural philosophy does not count as *scientia* according to Hobbes; a complete explanation should also show why Hobbesian natural philosophy is more epistemically certain than *cognitio*-based prudence. While Raylor is correct that Hobbesian natural philosophy is knowledge of the appearances, and also usefully draws attention to the discipline of meteorology, which for many since Aristotle was treated as a subalternate/subordinate/mixed science, Raylor's view neglects that it is because the principles of mathematics, which as *scientia* are epistemically certain, are borrowed for use within natural-philosophical explanations that Hobbes accords a greater epistemic weight to its conclusions; according to my account, the claims of natural philosophy are, thus, epistemically between mere prudence and *scientia*.

[4] Additional instances of Hobbes borrowing principles from geometry in natural philosophy beyond those I discuss in this chapter include the following: *De corpore* XXVI.6 (OL I.349; EW I.428), XXVI.8

concludes by contrasting the mixed-mathematics account I offer here with Shapin and Schaffer's view that Robert Boyle sought to defend experimental philosophy against the Hobbesian "beast of deductivism" (Shapin and Schaffer 1985, 176).

5.1 Aristotle, Barrow, Hobbes, and Wallis on Mixed Mathematics

5.1.1 Aristotle and Isaac Barrow on Mixed Mathematics

In *Posterior Analytics* I, Aristotle argues that "it is not possible to prove a fact by passing from one genus to another, e.g., to prove a geometrical proposition by arithmetic" (75a38–39).[5] For Aristotle, one cannot "prove by any other science the theorems of a different one, except such as are so related to one another that the one is under the other—e.g. optics to geometry and harmonics to arithmetic" (*APo* I.7, 75b14–17). Aristotle argues later that for sciences such as optics the 'that' will come from one science while the 'why' will come from a science which is "above" it (*APo* I.9, 76a4–13). In optics one may borrow geometrical principles because they study the objects of optics *qua* line and not *qua* object of sight (*Metaph* M.3 1078a14–16). In treating the objects of optics *qua* line, one treats a natural object as a mathematical object.

There has been some debate regarding the status of mathematical objects for Aristotle, given this account of mixed mathematics. Whereas Lear understands them as fictional objects (1982), Lennox views them as resulting from "taking a delimited cognitive stance toward an object" (1986, 37). In other words, one considers an object in a certain way. This is akin to how Hobbes describes mathematical objects, and as I have argued in chapter 3, Hobbes characterizes this cognitive ability as considering as.

Hobbes's younger contemporary Isaac Barrow appeals to and revises Aristotle's account of mixed mathematics in his *Mathematical Lectures* (1685). It is worthwhile to compare Barrow's view to Hobbes's because of their similar outlook in mathematics, especially since both held, against John Wallis, that geometry had priority over arithmetic (Jesseph 1993). In Lecture II, Barrow criticizes Aristotle and Plato for having distinguished pure from mixed mathematics by assuming that there are two kinds of things: intelligible things, the subject of pure mathematics,

---

(OL I.353; EW I.433–434), and XXVI.10 (OL I.357; EW I.438). Each of these explanations borrows geometrical principles related to circular motion from *De corpore* XXI (they cite XXI.10, XXI.11, and XXI.4, respectively). Hobbes similarly borrows geometrical principles from *De corpore* XXII.6 and *De corpore* XXIV.2 in optics in *De homine* 2.2 (OL II.8) (for discussion, see Adams 2014b, 39–40).

[5] See also *Physics* II.2 and *Metaphysics* M.1–3 (esp. 1078a14–17). For discussion, see Hankinson (2005), Lennox (1986), McKirahan (1978), and Wallace (1991).

132 HOBBES'S TWO SCIENCES

and sensible things, the subject of mixed mathematics (Mahoney 1990, 185). Barrow argues that "there exists in fact no other quantity different from that which is called magnitude, or continuous quantity, and, further, it alone is rightly to be counted the object of mathematics" (Barrow 1685, 39; trans. Mahoney 1990, 186). Since "magnitude is the common affection of all physical things," there is "no part of natural science which is not able to claim for itself the title of 'Mathematical'" (Barrow 1685, 40).

Some have taken Barrow's criticisms of the pure/mixed distinction as a rejection of mixed mathematics.[6] However, one might instead view Barrow's criticisms as a broadening of the purview of mixed mathematics (Malet 1997, 280ff). Indeed, Barrow continues in *Mathematical Lectures* to describe what will be the *new* mixed-mathematical disciplines, if his account is correct. In a way that resonates with Hobbes's comments from *De homine* X.5, discussed below, Barrow articulates the properly understood relationship between geometry and physics as follows: "[. . .] to return to Physics, I say there is no Part of this which does not imply Quantity, or to which geometrical Theorems may not be applied, and consequently which is not some Way dependent on Geometry" (Barrow 1734, 22; 1685, 41). As support for broadening mixed mathematics beyond the normally included disciplines, such as optics or harmonics, Barrow favorably mentions Aristotle's claim in *APo* (79a13–16) that "the physician chooses the cause from Geometry" when explaining why circular wounds heal more slowly (Barrow 1685, 40).

Seeing Barrow as broadening the purview of mixed mathematics connects Barrow to Hobbes, but there are important differences from Aristotle for both. For example, Barrow and Hobbes include motion in geometry (Mancuso 1996, 94ff), something which for Aristotle must be kept separate from mathematics (*Phys* II.2, 193b.35). The incorporation of motion into geometry makes kinds of motion themselves the *subject* of geometry, as, for example, in Lecture II of Barrow's *Lectiones Geometricae* (1670) and in Hobbes's discussions of motion in *De corpore* Part III, which part is entitled "Proportions of Motions and Magnitudes" (e.g., fermentation as a kind of circular motion considered in geometry is discussed below). Hobbesian mathematical principles also depart from the Aristotelian model, and from Barrow's model, because it is their status as maker's knowledge that makes them instances of *scientia*.[7]

---

[6] Mahoney (1990, 186). Similarly, Jesseph argues that Hobbes rejects the distinction between pure and mixed mathematics since Hobbes understands "body as the fundamental object of mathematics" (1999, 74–76). Nevertheless, Hobbes himself describes pure mathematics as that which treats quantities in the abstract (*in abstracto*), which is how he articulates the project of *De corpore* Part III, and takes "true physics" to be part of mixed mathematics (discussed below).

[7] Recent work has focused on other Aristotelian aspects of Hobbes's natural philosophy, in particular Leijenhorst (2002). See also Leijenhorst (1996).

5.1.2 Hobbes on First Philosophy, Geometry, and Mixed Mathematics

As discussed in chapter 3, in articulating the nature of *scientia* Hobbes appeals to the distinction between a demonstration of the 'that' (τοῦ ὅτι) and a demonstration of the 'why' (τοῦ διότι):[8] "We are said to know [*scire*] some effect when we know what its causes are, in what subject they are, in what subject they introduce the effect and how they do it. Therefore, this is the knowledge [*scientia*] τοῦ διότι or of causes. All other knowledge [*cognitio*], which is called τοῦ ὅτι, is either sense experience or imagination remaining in sense experience or memory" (OL I.59; EW I.66; Hobbes, 1981, 287–289). As Hobbes makes clear, causal knowledge is available only to *makers*—we make figures in geometry, which accords a special epistemic status to the geometrical principles we borrow in mixed-mathematical explanations (more on this below).

However, Hobbes's emphasis upon causal knowledge does not make the 'that' unimportant, as will become clear in this chapter and in chapter 6. On the contrary, although Hobbes excludes history from philosophy in *De corpore* I.8, he admits that both natural history and political history are "[. . .] very useful (no, indeed *necessary*) for philosophy [. . .]" (OL I.9; EW I.10–11; Hobbes 1981, 189; emphasis added). Similarly, in *De homine* XI.10 he claims that "[. . .] histories are particularly useful, for they supply the experiences/experiments [*experimenta*] on which the sciences of the causes rest" (OL II.100). Natural philosophers must know the 'that' from natural or political history, or from their own sense experiences or experiments.

Hobbes's criticism that Euclid "maketh not," and so his principles "ought not to be numbered among the principles of geometry" (EW VII.184; see also EW VII.202), implies that without providing a definition that specifies the mechanical procedure for constructing a figure, one cannot have causal knowledge about that kind of figure. Hobbes's view may seem intuitive for simple geometrical figures like lines and squares, but *De corpore* Part III moves beyond such figures and treats topics as far-ranging as endeavor and refraction. Thus, before discussing the geometrical principles that Hobbes borrows for the explanations to be discussed below, it is necessary to connect Hobbes's simple geometry of points and lines to Part III. For Hobbes, Part III concerns geometry because it treats *both* motion and magnitude, which are "the most common accidents of bodies" (OL I.75; EW I.68–69). At the end of Part III, he advises that "[. . .] we have considered motion and magnitude in themselves and in the abstract" (OL I.314; EW I.386). Part IV then treats phenomena of nature and concerns the "motion and magnitude of the bodies of the world, or which themselves exist in reality" (OL I.314; EW I.386).

---

[8] Some connection has been made in the literature between Zabarella and Hobbes, but there are significant differences between the two (see Dear 1988, 150–153; Hattab 2014; 2021).

134   HOBBES'S TWO SCIENCES

This distinction between 1) the features of bodies in the real world and 2) the abstract features of bodies, such as motion and magnitude, is essential for understanding Hobbesian geometry and mixed mathematics. Like others, Hobbes distinguishes between pure and mixed mathematics, but he does so in a way that does not fall prey to Barrow's criticisms of that distinction; sciences in which we discover abstract (causal) principles are pure, and sciences in which we borrow these principles for explanations are mixed, as in *De homine* X.5:

> [...] since one cannot proceed in reasoning about natural things that are brought about by motion from the effects to the causes without a knowledge of those things that follow from that kind of *motion*; and since one cannot proceed to the consequences of motions without a knowledge of *quantity*, which is geometry; nothing can be demonstrated by physics without something also being demonstrated *a priori*. Therefore physics (I mean true physics) [*vera physica*], that depends on geometry, is usually numbered among the mixed mathematics [*mathematicas mixtas*]. [...] Therefore those mathematics are pure which (like geometry and arithmetic) revolve around quantities in the abstract [*in abstracto*] so that work [in them] requires no knowledge of the subject; those mathematics are mixed, in truth, which in their reasoning some quality of the subject is also considered, as is the case with astronomy, music, physics, and the parts of physics that can vary on account of the variety of species and the parts of the universe. (Hobbes 1994b, 42; OL II.93)[9]

Immediately preceding this quotation, Hobbes identifies geometry as a form of maker's knowledge; as discussed in chapter 3, in this text and also in *Six Lessons* (EW VII.184), Hobbes argues that we possess maker's knowledge in geometry because "we ourselves draw the lines." However, since "the causes of natural things are not in our power" we can demonstrate only what their causes *may* be. Only God has access to the causes of natural things, but humans have access to the causes of things that they make, like geometrical figures and commonwealths.

The identification of geometry as maker's knowledge informs how we should understand Hobbes's claims about "true physics" in the extended quotation above from *De homine* X.5. To reason from the effects to possible causes in natural philosophy, one must know already what the causes may be. If we understand *a priori* to mean something like "from the causes" for Hobbes, then we are able to demonstrate "from the causes" when prior to a natural-philosophical investigation we already possess geometrical causal principles.

Even though BODY (POINT) is the fundamental object of mathematics, we find in this quotation from *De homine* X.5 that Hobbes nevertheless divides

---

[9] I have modified Gert's (Hobbes 1994b) translation.

mathematics into pure and mixed (see fn. 6). Pure mathematics treats quantities in the abstract, but natural philosophy also considers the qualities that "vary on account of the variety of species and the parts of the universe" (Hobbes 1994b, 42; OL II.93). So for Hobbes "true physics" "depends on geometry" and, since it also must consider qualities unique to certain kinds of bodies, it is a kind of mixed mathematics—experiences or experiments are mixed with geometrical principles.

Hobbes similarly distinguishes between pure and mixed mathematics in *Anti-White*. He claims that only arithmetic and geometry are presently mathematical because no one has written anything in morals or physics that is not "open to question." He asserts that in *Anti-White* I.1 that "[. . .] all the sciences would have been mathematical had not their authors asserted more than they were able to prove; indeed, it is because of the temerity and the ignorance of writers on physics and morals that geometry and arithmetic are the only mathematical ones" (Hobbes 1973, 106; Hobbes 1976, 24).[10] Hobbes argues that, in addition to geometry and arithmetic, mixed mathematics should also "be counted among mathematical" sciences (Hobbes 1973, 106; Hobbes 1976, 24). Among mixed mathematics, Hobbes includes astronomy, mechanics, optics, and music and leaves the door open for "others yet untouched" (Hobbes 1973, 106; Hobbes 1976, 24). Hobbes's idea that mixed mathematics must be broadened in this way, even to disciplines not yet existing, resonates with Barrow's assertion that even disciplines as disparate as medicine, politics, and zoology should be seen as dependent upon geometry and thus as mixed mathematics (Barrow 1734, 21–22).

The account in *Anti-White* emphasizes that mixed mathematics should be counted as part of mathematics since they consider "quantity and number, not [merely] abstractly [*non abstracte*], but with regard to the motion of the stars, or the motion of heavy [bodies], or with regard to the action of shining [bodies], and of those which produce sound [. . .]" (Hobbes 1973, 106; Hobbes 1976, 24–25). So in addition to considering quantity and number, which pure mathematics does, mixed mathematics treats the unique qualities that particular bodies possess.

Hobbes's claim that in pure mathematics we treat a body in the abstract differs from seeing abstraction as grounding mathematics. Hobbes, like Aristotle, is licensed in borrowing the 'why' from geometry because one treats a natural body like a mathematical object; as articulated in chapter 3, he uses variations of the phrase "consider as" to refer to this cognitive activity. For example, Hobbes's difficulty with Euclid's definition of 'point' as "a breadthless length" is that "there is no such thing as a broad length" (EW VII.202).[11] Instead, Hobbes argues that a line is "a body whose length is considered without its breadth" (EW VII.202). Similarly, we consider bodies as points, like when we call the Earth a point when

---

[10] I have amended Jones's translation (Hobbes 1976) to reflect Hobbes's use of *moralis*.

[11] Hobbes's criticisms of Euclid's account of mathematical definitions apply to Wallis as well (OL IV.41–42).

136 HOBBES'S TWO SCIENCES

discussing its annual revolution (OL I.98–99; EW I.111). Thus discussions about lines in geometry refer to *bodies* considered without breadth as mathematical objects, though as bodies they actually do have breadth. This way of characterizing Hobbesian mathematical objects has affinity with Lennox's understanding of Aristotelian abstraction as "taking a delimited cognitive stance toward an object" (Lennox 1986, 37). In other words, for Aristotle one considers an object in a certain way, and the same can be said for Hobbes. However, whereas Aristotle arrives at mathematical principles by abstracting away physical features, Hobbes arrives at mathematical principles by an analysis of complex conceptions received in experience down to the "simplest conceptions" (OL I.62; EW I.70) and then by synthetically constructing geometrical figures and providing definitions for them using those simplest conceptions. For example, we analyze the complex conception SQUARE and terminate in the simples BODY, PLACE, and MOTION. Putting SQUARE back together provides knowledge of the "cause of the square" (Hobbes 1981, 293; OL I.61; EW I.69).[12] Why would humans engage in a challenging project like natural philosophy when it requires both difficult *a priori* work in geometry along with careful observations from experience? Hobbes's answer, discussed in chapter 3, is that few individuals will be able to do so because of the unique situation that is required for it to be possible: individuals must be driven by curiosity, as humans are, but they also must have leisure and a method not only for developing geometry but also for mixing causal principles from geometry with observations from sense experience.

De corpore Part II ("The First Grounds of Philosophy") also plays a role within explanations in natural philosophy. In chapter 4, I argued that the annihilatory thought experiment in *De corpore* VII–VIII articulates definitions by explication for conceptions like BODY and MOTION, and then these definitions serve as the starting point for constructions within the geometry of *De corpore* Part III. However, in addition to this role, Hobbes argues in first philosophy for two *a priori* principles: one related to motion/rest in *De corpore* VIII.19 and another, derived from it, requiring that all motion in a body must be caused by the motion of another contiguous, moving body in *De corpore* IX.17. The arguments that Hobbes offers to support these *a priori* principles—I will adopt Jesseph's terminology and call these the "persistence principle" and the "principle of action by contact" (2006, 131–135)—are within the framework of the annihilatory thought experiment begun in chapter VII because the reader is asked to "[. . .] suppose that some finite body exist and be at rest, and that all space besides be empty [. . .]" (OL I.102; EW I.115). Hobbes's argument for the persistence principle concludes that it is not possible even to imagine a body self-initiating motion or rest. He argues that a body in empty space would have the same reason to move one direction as any other and

---

[12] This example is discussed in detail in chapter 3.

thus, were it possible for a body to self-move, this would result in the body being "moved alike in all ways at once" (OL I.102; EW I.115). Likewise, if it were possible to imagine a body self-initiating its own rest, then equally at each moment in time the lone body would have a sufficient reason to rest; this, Hobbes argues, is "not intelligible" (OL I.103; EW I.116). The persistence principle plays a direct role in Hobbes's criticisms of Boyle and others concerning the motions of bodies like crossbows and wool (discussed below in subsection 5.3.1)—it rules out any explanation that relies upon self-motion.

In *De corpore* IX.7, Hobbes derives the principle of action by contact by citing the first principle and then imagining two bodies. The persistence principle shows, Hobbes claims, that any change in motion or rest must be from some external body, but a body in empty space will persist in whatever state it is (OL I.110; EW I.124). Hobbes's reasoning is that, given the persistence principle, if there is *nothing* around a resting or moving body, it simply cannot change. It follows that any change initiated in a body by another body must happen only insofar as that second body is "external" to the first, i.e., where external is taken to mean contiguous. Agreeing with Jesseph's analysis (2006, 135), Hobbes has simply declared action at a distance incoherent in what could be taken as a question-begging move. Contiguity alone will be insufficient for one body to cause change in another, since if two resting bodies are contiguous, they will remain at rest until something else disturbs them (from the persistence principle), so Hobbes draws upon his definition of 'cause' as the aggregate of all of the accidents in a given system. For an aggregation to result in the motion of a body initially at rest, that body must not only be touched by some other contiguous body but one that is also moving. Hobbes uses—and explicitly cites—the principle of action by contact within the explanation of sense in *De corpore* XXV (discussed below in subsection 5.2.1).

While Hobbes thought that borrowed principles from geometry could serve as the possible causes of natural phenomena, his use of these two *a priori* principles from first philosophy functioned to limit what could serve as a possible cause.[13] While Hobbes admits that humans can never know the actual causes of natural phenomena, given their *a priori* status these two principles immediately rule out any appeal to self-initiated motion or rest (change) as well as action at a distance. This function of these principles is clear in Hobbes's repeated criticisms of the doctrine of rarefaction and condensation, which doctrine sought to explain the apparent change of magnitude of a body like a sponge that is wetted and then after being compressed seems to have a smaller magnitude than before being

---

[13] Hobbes uses other principles from first philosophy that I will not discuss in this same way (e.g., a principle related to the division of bodies and places used in *De corpore* XXV.6 [OL I.321; EW I.394–395] and another principle related to the necessity of an effect following from a necessary cause used in *De corpore* XXV.13 when explaining deliberation [OL I.333; EW I.409]).

138   HOBBES'S TWO SCIENCES

dampened.[14] As Jesseph argues, Hobbes held that this doctrine was "incoherent because it violates the persistence principle and the principle of action by contact" (2016, 144) insofar as it appeals to a body's supposed ability to expand and compress itself without an external body causing those changes. Just as Hobbes viewed this doctrine as inconceivable because it was in violation of these principles (see discussion in Jesseph 2016, 144–145), he similarly argues that Boyle's hypothesis— recounted through speaker B in *Dialogus Physicus*—that air moves by an elastic force is inconceivable because it would require considering air as capable of self-motion (discussed below in subsection 5.3.1).

5.1.3 Hobbes against Wallis on Mixed Mathematics

While Hobbes's understanding of mixed mathematics has commonalities to Aristotle's and Barrow's, his views contrast sharply with those of his immediate contemporary John Wallis. As discussed in chapter 1, Hobbes and Wallis disagreed concerning the fundamental nature of the objects of mathematics. Wallis held that pure mathematics treats quantity absolutely, that is, quantity absolutely as "abstracted [*abstrahitur*]" from matter. In mixed-mathematics disciplines, Wallis held that the particular subject in which quantity exists is "connoted [*connotatur*]" alongside the "consideration" of quantity (1657, 2–3). Hobbes's view grounded mathematical objects in body, and thus any accident of a body, whether geometrical or not, was examined in the mind by "considering as." In *Examinatio*, Hobbes ridicules Wallis's claim that somehow "connotation" of a subject differs from "considering" quantity, saying that Wallis "writes what he learned as a boy" (OL IV.24).

Hobbes's other complaints against Wallis's understanding of the pure/mixed distinction relate to their roles in explanations within mixed-mathematics disciplines. In chapter 1 of *Mathesis Universalis*, Wallis discusses various examples of mixed-mathematics disciplines and holds that mixture is gradational "according as they approach pure mathematics more or less" (1657, 3). For example, Wallis suggests that in astronomy some aspects are drawn from pure mathematics while many other aspects are simply physical and are "observed from natural history" (1657, 3). So far Hobbes would seem to agree with this generic description of mixing observations from experience with principles from geometry. However, Hobbes raises worries concerning Wallis's claim that "Astronomy teaches [*docet*] that the Equator and the Zodiac intersect each other in two points [. . .]" (Wallis 1657, 3). Hobbes responds in *Examinatio* by assigning disciplinary priority

---

[14] For example, see Boyle's discussion of various accounts of the behavior of sponges when defending his experiments in "A Defense of the Doctrine Touching the Spring and Weight of the Air" (1682, 34–46).

to the geometer: "Astronomy does not teach [*docet*] this, nor is it the task of an Astronomer to demonstrate [*demonstrare*] it, but [it is the task] of a Geometer" (OL IV.25).

At first glance, it may seem Hobbes is simply arguing about terminology and reasserting his view that 'demonstration' and 'teaching' are reserved only for *scientia* like geometry (discussed in chapter 4). After all, on Wallis's account, the explanation for the Earth's motion will involve both observations related to its annual path along with principles from pure mathematics, such as principles concerning the nature of a circle and the intersection of circles (Wallis 1657, 3). Additionally, Wallis's account adds that "mathematical principles and instruments [*Mathematicis tam Principiis quam Instrumentis*]" aid in making these observations. Hobbes rejects the view that measurement plays a role in the observations for astronomical explanations and states only that the "Astronomer observes two motions of the sun, Diurnal and Annual, occurring in two large circles, and then, as a Geometer [*ut Geometra*], investigates the angle that they make" (OL IV.25).

Hobbes's reservation about using the terms 'demonstrate' and 'teach' for astronomy and other mixed disciplines makes sense considering his broader view that natural philosophers can offer only possible causes. Additionally, Hobbes's criticism seems to stem from Wallis's way of thinking about *why* mathematics can be used when explaining natural-philosophical phenomena. Whereas Hobbes holds that mathematical objects are drawn from "considering as" bodies, but nevertheless human makers can gain *scientia* from constructions using them, Wallis's account of why using mathematics is appropriate in mixed disciplines is accidental: "Although indeed all these things [e.g., motion, weight, and time] are treated in the mathematical disciplines, yet not in and of themselves, and not primarily, but insofar as they are either measured or counted." These other disciplines—the mixed-mathematics disciplines—"pertain to Mathematics" only insofar as "they are (reductively) capable of numbers or measures" (1657, 2). In Wallis's view, mixed-mathematics disciplines are only instrumentally *mathematical* to the extent that it is appropriate to ask how many or how much of something there is. In contrast, Hobbes viewed mathematics as providing human makers with causal knowledge (*scientia*), and principles from mathematics could be used in natural-philosophical explanations because they were drawn from considerations of physical bodies themselves.

Finally, Hobbes disagreed with Wallis's claim that demonstrations in pure mathematics were of three types: *deductio ad absurdam*, ostensive demonstration τοῦ ὅτι ('that'), or ostensive demonstration τοῦ διότι ('why').[15] Wallis claimed not only that the demonstration 'that' had a role in mathematics but also that ostensive τοῦ διότι demonstration—what he called "the most perfect of all"—was one in which

---

[15] For discussion of Wallis's view of these types of demonstration, see Rampelt (2019, 192–193).

140 HOBBES'S TWO SCIENCES

both 'that' and 'why' were shown in the same demonstration. An example of such an ostensive τοῦ διότι demonstration he provides is

> [...] if someone demonstrates that all radii of the same circle are equal from the fact that the circle is defined (or at least can be defined) as a plane figure contained within one curve that is everywhere equidistant from the middle of the space it comprehends. For if the essence of a circle postulates that its periphery is equidistant from the center, it follows immediately as from a true and proximate cause that all radii (by which this distance is measured) are also equal. (1657, 13–14; Wallis OM 1.23–24; trans. in Jesseph 1999, 204–205 fn. 11)

Jesseph rightly notes that Hobbes downplays the significance that Wallis and others attributed to the distinction between τοῦ ὅτι and τοῦ διότι in pure mathematics, in this case geometry (1999, 204). Indeed, both interlocutors in Hobbes's *Examinatio* think Wallis's claim to find τοῦ ὅτι demonstrations in mathematics worthy of ridicule. Speaker B mocks that each time he hears Wallis say "*Demonstratio* τοῦ ὅτι" he has to restrain himself to avoid asking "τοῦ ὅτι τί [*why* is the that?]" (OL IV.42). B continues that it makes little sense (in mathematics) to hold that there could be a demonstration 'that' because "we do not know that a thing is unless we know why it is through knowing [*scire*] the cause" (OL IV.42). Speaker A concurs and summarizes Aristotle's understanding (shared by Hobbes, as discussed above) that seeking causes through known (*cognitos*) effects cannot be done accurately (*accurate*) "because similar effects do not always and necessarily have similar causes." As a result, interlocutor A notes that Aristotle viewed demonstrating an effect through its cause as stronger, while holding that τοῦ ὅτι failed to be scientific knowledge (*scientifica*) and did not provide "a true demonstration" (OL IV.42). However, these criticisms should not be seen as Hobbes giving up on the distinction between τοῦ ὅτι and τοῦ διότι wholesale (*pace* Rampelt 2019, 194). Hobbes clearly rejects the idea that non-causal τοῦ ὅτι should play a role in geometry; were this the case, geometry would fail to be an instance of maker's knowledge. Nevertheless, in discussions and practice of mixed mathematics, Hobbes saw τοῦ ὅτι as mixing with causal principles from geometry in explanations of natural-philosophical phenomena.

5.2 Mixed-Mathematical Explanations in *De corpore*

5.2.1 Explaining Sense in *De corpore XXV*

De corpore Part IV follows the second path of philosophy, beginning from "known effects or phenomena," but since the actual causes of natural things are unavailable to us, Part IV shows how "they can be generated" (OL I.315–316; EW I.387–388). These comments reflect different levels of certainty for geometry and natural philosophy. The maker's knowledge possible in the constructions of geometry and civil philosophy is the paradigm for *scientia* because in these it is possible to know *actual* causes, but explanations in natural philosophy lie epistemically between 1) the certainty of geometry and civil philosophy and 2) the limited prudence characterizing those who rely solely on memory and associations. Borrowing maker's knowledge from geometry transfers some of the certainty had there to a natural-philosophical explanation, but only *suppositional certainty*: the natural philosopher can know that if such a motion *were* present in actual bodies, then certain effects would necessarily follow from that motion.[16]

Explaining sensation is necessary because Hobbesian natural philosophy starts from the appearances of nature. So Hobbes first discusses "appearing itself," which he calls the "most admirable" of all appearances (OL I.316; EW I.389). Since appearances are the starting points by which all other things are known, he argues that sense is the principle by which all other principles are known: "all knowledge may be said to be derived from [sense]" (OL I.316; EW I.389). Succinctly, we know only through phantasms, but the only way that we become aware of and inspect phantasms is by sense. Thus, an inquiry into phantasms must begin with sense.

Sense is a stopping point against any potential regress which might require a faculty beyond it to be supposed, such as a faculty of the intellect. We are aware of sense not by some other faculty, but by *sense itself* since anyone who has sensed remembers that he has sensed: "For sensing oneself [as] having sensed is to remember" (OL I.317; EW I.389). Hobbes explains sensation in *De corpore* XXV with three separate definitions of sense in a series of refinements, with each refinement due to borrowing a causal principle from either geometry or first philosophy. As will be discussed below, the first definition understands sense as "nothing other than motion of some of the parts inside the sentient, which moved parts are parts of the organs by which we sense" (OL I.317; EW I.390). I will show that this definition mixes an appeal to experience—something that can be "observed"—with a borrowed principle related to mutation and motion from first philosophy. Two further refinements to this initial definition are developed by borrowing additional principles, one from first philosophy and the other from geometry. I have summarized the steps Hobbes takes to reach the final definition as follows:

Hobbes's Explanation of Sense in *De corpore* XXV

1. All sense is mutation of the sentient body (known from experience; the 'that').
2. All mutation is motion or endeavour, and endeavour is also motion (borrowed principle from first philosophy; *De corpore* IX.9).

---

[16] Jesseph (2006, 139) makes a similar point regarding applying principles from first philosophy to natural philosophy.

142   HOBBES'S TWO SCIENCES

3. Therefore, all sense is motion in the sentient body (from 1 and 2).
4. The motion of any body A occurs only by means of some other body B which is contiguous to A and presses upon A (principle of action by contact, borrowed from first philosophy; *De corpore* IX.7).
5. Thus, all sense occurs by contiguous bodies pressing upon the sentient body, i.e., the organs of sense, which motion continues in the sentient body (from 3 and 4).

Remaining explanandum: why do humans perceive bodies as outside of them if sense can be explained solely in terms of contiguous bodies pressing against one another?

6. All resistance is endeavour contrary to another endeavor, i.e., reaction (borrowed principle from geometry, *De corpore* XV.2; the 'why').
7. Supposition: the body of the sentient has an internal endeavour outward which resists inward motion, such as motion from the objects of sense.
8. Thus, (*if* the sentient body behaves according to the supposition in 7) the resistance of the internal parts of the sentient's body against the inward motion from the object causes human perception of bodies as outside of them.

Hobbes appeals to everyday experience—the fact that the appearances of things continually change—to demonstrate the 'that':

[. . .] it is proper in the first place to observe that our phantasms are not always the same, but new ones are constantly being created and old ones are disappearing, just as the organs of sense are turned now to one, now to another object. Therefore, they are produced and pass away, from which it is understood that they are some mutation of the sentient body. (OL I.317; EW I.389)

Observing the phantasms that arise in experience, it is evident that with each new body we encounter, a new phantasm arises and disappears once we turn elsewhere. Since phantasms are continually produced and pass away, there must be mutation in the sentient body.

Hobbes next considers what follows from knowing that phantasms are mutation. He borrows a principle from first philosophy:

But that all mutation is something having been moved or endeavoured, (which endeavor [*conatus*] is also motion) in the internal parts of the thing changed has been shown (cap. 9., art. 9.) from this: that while the smallest parts of some body stay the same having been mutually positioned, nothing new happens to those parts, (unless perhaps it may be possible that every part may be moved at the same time), except that it both be and appear to be the same, which at first it was and appeared to be. (OL I.317; EW I.389–390)[17]

Adding this borrowed principle from *De corpore* IX.9 to the earlier demonstration of the 'that', Hobbes formulates the first definition: "[...] sensation in the sentient can be nothing other than motion of some of the parts inside the sentient, which moved parts are parts of the organs by which we sense" (OL I.317; EW I.390). The principle that "all mutation is something having been moved or endeavoured" implies that motion must be responsible for the mutation involved in sensation. Hobbes notes with this first definition that he has shown the "subject of sense" to be the organs of sense in which phantasms are created. He also claims to have discovered "part of its nature": it is "some internal motion in the sentient" (OL I.317–318; EW I.390).

Hobbes borrows a second principle from *De corpore* IX in the first refinement of this definition. He wants to show that the motion that causes sensation in the sentient must originate from the internal motions of the parts of the object of sense and be carried to the subject of sense. He does this by borrowing from first philosophy again: "[...] it has been shown (cap. 9., art 7.) that motion cannot be generated except by [a body] moved and contiguous. From which the immediate cause of sensation is understood to be in this, that it both touches and presses the first organ of sense" (OL I.318; EW I.390). He formulates the second definition: "[...] sense is some internal motion in the sentient, generated by some motion of the internal parts of the object, and propagated through media to the inmost parts of the organ" (OL I.318, EW I.391). This use of the principle of action by contact from first philosophy constrains what can count as a possible cause—only a moving body touching the organ of sense could be its immediate cause.

Thus far we have discussed steps 1–5 of Hobbes's definition. These *a priori* principles from first philosophy have their justification in being conceptual truths (Jesseph 2006, 130). As conceptual truths, Hobbes's claims in step 3 and step 5, are merely the application of the definition of 'mutation' and the principle of action by contact to what has been demonstrated from experience, i.e., that sensation is a mutation of the sentient body. With the last refinement, however, Hobbes does something different. He borrows a geometrical principle related to 'resistance' and must *suppose* that sentient bodies behave according to that causal principle.

Hobbes has "almost defined what sense may be" (OL I.318; EW I.391). It remains to explain why humans perceive objects of sensation as outside of them rather than inside of them. Given his account of phantasms as caused by *internal*

---

[17] The Molesworth *Latin Works* edition and the 1655 edition (cf. Hobbes 1655, 224) incorrectly record this citation as to *De corpore* VIII.9. Schuhmann (Hobbes 1999, 268) corrects the citation to *De corpore* IX.9.

144   HOBBES'S TWO SCIENCES

motion in the sentient, he has not yet explained why we do not perceive those objects as *inside of us*. Since Hobbes explains sensation by appeal only to the cause of motion, this problem is present for Hobbes in a way that it is not for others.[18] The motion from the object of sense that is transmitted by a medium and the motion from the reaction of a sentient's body are the only *explanantia* that are mechanically intelligible for Hobbes. To be consistent with this constraint, Hobbes posits that we perceive phantasms as caused by outside bodies because of outward motion from the resistance of our body against the inward motion. Hobbes borrows a causal principle for 'resistance' from *De corpore* XV.2 (Part III): "Likewise, it has been shown (cap. 15., art. 2.) that all resistance is the endeavour contrary to [another] endeavor, that is, reaction" (OL I.318; EW I.391). In sensation, this reaction occurs because "the natural internal motion of the organ itself" resists the motion from the object of sensation. Since the endeavour moves outward due to this resistance, the phantasm "always appears as something situated outside of the organ" of sense (OL I.318; EW I.391). Rather than being a conceptual truth from first philosophy, like the earlier borrowed principles, this principle of resistance from Part III is geometrical because it, like the concept of 'endeavor' (introduced in *De corpore* XV.2 and used in the definition of 'resistance'), is a kind of simple motion which can be treated according to proportions. For example, it is intelligible to compare the proportion of the velocities of two endeavors on Hobbes's account (OL I.178; EW I.206).

This borrowed causal principle allows Hobbes to formulate the final definition of sensation: "[...] a phantasm made by means of a reaction from an endeavour to [the] outside, which is generated by an internal endeavour from the object, and there remains for some time" (OL I.319; EW I.391). Since we are explaining the behavior of natural bodies, we cannot know with certainty that the body of the sentient *actually* reacts against the inward motion in this way. However, we can use the borrowed principle to form a supposition: *if* the body of the sentient resists the inward motion with an outward-directed endeavour, *then* the motion would continue until leaving the body. Hobbes describes the final definition of sense as "from the explication of its causes and its order of generation" (OL I.318–319; EW I.391). This definition is possible only by borrowing a causal principle from geometry (the 'why') and mixing it with something known by experience (the 'that').

---

[18] Consider the case of vision, though this problem applies to all of the senses equally for Hobbes. Hobbes's explanation of vision does not employ an image on the back of the retina or appeal to natural triangulation similar to "surveyors" (*mensorium*) as Kepler's does in chapter III, Propositio IX of *Ad Vitellionem Paralipomena* (1604, 63). Furthermore, Hobbes's explanation does not assume that our ability to know an object's location occurs "as if by natural geometry" like Descartes in *Dioptrique* (1637/2001, 104). Instead, Hobbes endeavors to explain all aspects of vision by motion alone.

5.2.2 Explaining Effects of Heat on Human Perceivers in *De corpore* XXVII

Prior to explaining light, heat, and color in *De corpore* XXVII, Hobbes introduces several suppositions. First, he supposes that no matter how small some bodies may be, he will "suppose" only that their size is not smaller than what the phenomena themselves require (OL I.364; EW I.447). Second, regarding the motion of the bodies under consideration, he supposes only what is needed for the "explication of [their] natural causes" (OL I.364; EW I.447–447). Finally, he supposes that "in the parts of pure ether" there is no motion except what is transferred "by the bodies floating in it" and that these parts of the ether are not liquid (OL I.364; EW I.448). To explain the cause of the light (*lux*) of the sun, Hobbes introduces an additional supposition: that the sun "by its simple circular motion" moves the parts of the ether that are near it (OL I.364; EW I.448).[19] Hobbes defines simple circular motion as follows: "[. . .] in simple circular motion it is necessary that every straight line taken in the moved body be always carried parallel to itself [. . .]" (OL I.259; EW I.318). By simple circular motion, Hobbes means something akin to a sieve-like motion around a center point. Take Circle A in Figure 5.1, which has contained within it the straight lines B and C.

If Circle A were simply to rotate around its center point, lines B and C would fail to remain parallel to their previous positions from T1 to T3. However, Hobbes argues that if a circle like Circle D moves like a sieve in simple circular motion around its center point, then lines B and C will always be parallel to their past positions when carried with Circle D. For ease of illustration, the dashed lines in Circle D represent past locations through the passage of time as it continued to move in simple circular motion.

Hobbes is aware that this type of motion, which serves as the foundation of much of his natural philosophy, is potentially unintuitive. In *De corpore* XXI.2, he compares it to if "many pens' points of equal length were fastened, [someone] might with this one motion write many lines at once" (OL I.261; EW I.320) (discussed more in subsection 5.3.2). In *Examinatio*, interlocutor B worries that Hobbes's readers will fail to understand what he means: "[. . .] although such motion is most suitable for producing nearly all the phenomena of nature, since it has not been observed and explained by anyone before him, few readers can easily follow the course of the discourse in which that motion is described and calculated" (OL IV.226–227). Speaker A provides yet another attempted analogy to make this easier to imagine: imagine a ball to which a writing stylus is attached which someone uses

---

[19] Galileo's *Dialogo* may be a source of Hobbes's use of simple circular motion (Baldin 2020; Brandt 1928, 330ff; Henry 2016; Mintz 1952). Simple circular motion is another connecting point between Hobbes and Barrow; see Barrow's *Geometrical Lectures* (Lecture II; 1670).

146 HOBBES'S TWO SCIENCES

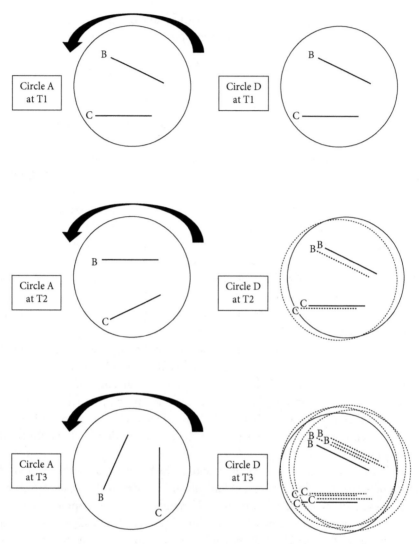

Figure 5.1 Rotation of a Circle versus Simple Circular Motion.

to write a letter of the alphabet using a continuous stroke. As the person writing with the stylus (or even if there were many styluses attached to the ball) moves in continuous motions to write letters, speaker A argues that there will be simple circular motion present "[...] when the points of the ball describe circles but also when those points describe any other figures, provided that those points return to their starting positions by their own motion" (OL IV.227). As A notes, this type of motion is such "that any line taken on the ball is always carried parallel to itself" (OL IV.227).

Returning to *De corpore* XXVII, as the sun moves in simple circular motion (as supposed), it likewise moves the parts of the ether around it. This motion propagates through the medium, reaching the organ of sense and the heart of the perceiving human. Referring back to his explanation of sensation in *De corpore* XXV, Hobbes states that the endeavour outward is "called light [*lumen*] or the phantasm of a lucid [*lucidi*] [body]" (OL I.365; EW I.448). These considerations provide the possible cause of the light of the sun (*lucis solaris*). Thus far in chapter XXVII, Hobbes seems to have asserted mysteriously that the endeavour moving outward from a perceiver's body is the cause of the light of the sun. Hobbes is, of course, using the vocabulary relevant to the distinction between *lux* and *lumen* when he provides this possible cause of the *lux* of the sun.[20] However, Hobbes uses this vocabulary to differentiate between two motions that he posits as causing the perception of light: 1) the motion from the luminous body and 2) the resistance against that motion by the sentient body. The *lux* of the sun is its simple circular motion propagated through media, and *lumen* is created because of the outward reaction of perceivers' bodies to *lux*.

Hobbes's account in *Elements of Law* helps clarify this point. Instead of the simple circular motion that we find in *De corpore*, in *Elements* Hobbes posits that the motion produced by fire and other lucid bodies is "dilation, and contraction of it self alternately, commonly called scintillation or glowing" (EL I.II.8). Apart from this difference in the type of motion between *Elements* and *De corpore*, though, his account of the perception of light is largely the same:[21]

> Now the interiour coat of the Eye is nothing else but a piece of the *Optick* nerve; and therefore the motion [from the lucid body] is still continued thereby into the *Brain*, and by *resistance* or reaction of the Brain, is also a *rebound* into the Optick nerve again; which we *not conceiving* as motion or rebound from within, do think it *without*, and call it *light* [...]. (EL I.II.8)

Thus, *lux* from the sun is nothing other than simple circular motion coming from the lucid body (or scintillation in *Elements of Law*). This motion propagates through media, continues to the eye, and then rebounds outward after meeting resistance, creating *lumen*, which causes the phantasm of light.

Following this explanation of the possible cause of the *lux* of the sun, Hobbes focuses on the felt heat that accompanies the light of the sun. This explanation occurs in three steps, two of which are appeals to experience. The first appeal to

---

[20] Lindberg (1978, 356) discusses the entrenchment of the *lux*/*lumen* distinction through the Latin translation of Avicenna's *De Anima*. For Avicenna *lux* referred to the light from luminous bodies, such as the sun, and *lumen* referred to the effect of *lux* upon the medium and the non-luminous bodies which it lit. For discussion of disagreement between Hobbes and Margaret Cavendish on the *lux*/*lumen* distinction, see Adams (2022).

[21] The terms *lux* and *lumen* are not present, of course, since *Elements* was composed in English.

148 HOBBES'S TWO SCIENCES

experience is used to differentiate Hobbes's intended explanandum from another, and the second is to establish the 'that'. In the third step, he borrows a principle from *De corpore* XXI.5 related to a type of circular motion that he calls 'fermentation'. Like Barrow does in his *Geometrical Lectures* (see fn. 19), Hobbes treats simple circular motion within geometry. According to Hobbes, fermentation is a type of simple circular motion wherein bodies perpetually change place (the status of simple circular motion as *mathematical* is the focus of subsection 5.3.2 below; for now, I will discuss the appeal to it within the present explanation).

Hobbes's first appeal to experience shows what may be inferred about lucid bodies from the heat they cause in us. We know by experience what it is to perceive heat in ourselves when we grow warm, but we know what it is "in other things by ratiocination" (OL I.365; EW I.448–449). He distinguishes between 1) the sensation of heat and 2) what we can know about the things that produce heat: "we recognize fire or the sun making warm, but we do not recognize that it may be hot" (OL I.365; EW I.449). Although in the case of heat that is caused in us by other creatures we know that those creatures are themselves hot, like heat caused by a dog lying on one's lap, Hobbes argues that we cannot make the same inference from the heat caused in us by the sun to the properties of the sun itself. The inference Hobbes is opposing would claim something like the following: anytime a creature causes warmth in another body, that creature itself is warm; thus, when a body like the sun causes warmth, it must be warm. Hobbes thinks that we can no more assert this than we can say that "fire causes pain, therefore [fire] is in pain" (OL I.365; EW I.449).

Hobbes next makes a second appeal to the everyday experience of being warm to establish the 'that':

> [. . .] when we are growing hot, we learn that the spirits, blood, and whatever is fluid in our bodies is called forth from the interior parts to exterior as the degree of heat is more or less, and the skin swells up. (OL I.365; EW I.449)

Hobbes focuses upon a feature of the experience of being warm—the pores sweat and the skin swells. This appeal to experience demonstrates that this sweating and swelling occurs when a body is heated.

Next Hobbes provides the 'why' for the skin's swelling and the possible "cause of the heat of the sun." Hobbes's stated explanandum is the cause of the heat of the sun, but given his reservations on the inference that can be made from 'heat of the sun', his actual explanandum is the cause of the *sensation of heat* from the light of the sun. To provide this cause, Hobbes borrows a principle related to fermentation from *De corpore* XXI.5 (fermentation will be discussed more below in section 5.3 related to the nature of air).[22]

---

[22] Hobbes claims that in *De corpore* XXI he has explained how the air is moved by the "simple circular motion of the sun." However, as Schuhmann notes (Hobbes 1999, 304 fn. 2), this likely refers to

Hobbes first supposes that the sun's simple circular motion moves the air around it so that the parts of the air "perpetually change their places with one another" (OL I.366; EW I.449–450). This motion is propagated from the sun to the air that surrounds humans. Hobbes identifies this perpetual change of place with the process of fermentation, drawing attention to an earlier demonstration in *De corpore* where he explains how water is drawn up into the clouds by the same cause of the circular motion of fermentation.[23] Like water that forms clouds when drawn from the ocean, Hobbes explains how "from our bodies the fluid parts from the insides to the outsides may be drawn out by the same fermentation" (OL I.366; EW I.450).

Hobbes is drawing on common knowledge that fermentation causes heat, but he is describing this common notion in terms of a particular kind of simple motion. Fermentation is a type of circular motion that involves the perpetual change of place by the parts of air that results in the joining together of homogeneous parts and the production of heat (*De corpore* XXI.5, OL I.263–265; EW I.323–325). His interest in this type of motion is to show how, as a type of simple circular motion, we can use fermentation to explain how air draws water into the clouds and human sweating. This is a peculiar usage of fermentation, to be sure, but Hobbes is taking the abstract geometrical account of fermentation as this type of motion and using it as a cause in explanations of phenomena as diverse as sweating in human bodies and cloud formation over the ocean, both of which are mixed explanations that take into account the "qualities of the subject," as described in *De homine* X.5.

---

an earlier version of *De corpore* since there is no discussion of the simple circular motion of the sun in XXI.5. Although this claim is absent from extant versions of *De corpore* XXI.5, for my purposes it is sufficient that Hobbes does introduce the concept of fermentation in that article, and that fermentation provides the causal principle for the explanation of the possible cause of the heat of the sun. Furthermore, Hobbes's supposition that the sun moves the air around it by simple circular motion occurs throughout *De corpore* (OL I.351; EW I.430; OL I.358; EW I.440; OL I.364; EW I.448; OL I.367–368; EW I.451–453; OL I.381; EW I.468).

[23] There is a discrepancy between extant versions of *De corpore* related to this reference to cloud formation and fermentation. Both Latin editions of *De corpore* (1655; 1668) and also the Molesworth *English Works* and English edition of 1656 record this as a reference to a demonstration contained in *De corpore* XXVI.8, but such a demonstration is absent from XXVI.8. However, this reference is missing from the Molesworth *Latin Works* edition, and Schuhmann follows OL, claiming that this demonstration Hobbes cites might have been in an earlier version of *De corpore* XXI.11 (cf. Hobbes 1999, 305). Schuhmann may have found evidence for this claim regarding *De corpore* XXI.11 (though he does not state this) because of a later reference in *De corpore* XXVIII.14 to the formation of clouds. There the 1655 edition (Hobbes 1655, 276) and the Molesworth *Latin Works* edition (OL I.391) cite an explanation of the formation of clouds that is also supposed to be in *De corpore* XXI; Schuhmann (Hobbes 1999, 323) explains the fact that such an explanation is missing in *De corpore* XXI by supposing that it might have been present in an earlier version of chapter XXI. On this citation to chapter XXI, the *English Works* edition (EW I.480) follows the English edition of 1656 (Hobbes 1656b, 357), recording this citation as being to *De corpore* XXVI instead. In positing an earlier version of *De corpore* XXI as the likely location of this explanation, Schuhmann neglects the possibility that this explanation of the formation of clouds due to fermentation was moved to *De corpore* XXVIII.2, where Hobbes *does* discuss the formation of clouds (OL I.381; EW I.468–469). Perhaps Hobbes had once included this explanation in *De corpore* XXI (Part III) as an *example* of how the cause of fermentation as a type of circular motion could be used as the 'why' in natural philosophy, but later he moved the example so that it was where one would expect it—within the section on natural philosophy (Part IV).

150   HOBBES'S TWO SCIENCES

The account of fermentation in *De corpore* XXI.5 is geometrical not merely because it is within Part III, but more importantly because it describes a type of simple circular motion irrespective of the unique qualities of particular bodies like human bodies or rain clouds. As a type of simple motion, Hobbes, like Barrow, holds that it is the work of geometry.

The status of fermentation as simple circular motion, and thus as mathematical, is a focus of subsection 5.3.2 below. For now, here is how Hobbes puts it to use. When parts of air that are contiguous to the body of an animal ferment by perpetually changing places with one another, "the parts of the animal contiguous to the medium may endeavour to enter into the spaces of the divided parts" (OL I.366; EW I.450). Therefore, the "most fluid and separable" parts of the animal go out first, and their place is filled by other parts which are able to transpire through the pores of the skin.

What happens to the non-fluid parts of animal bodies that are not able to be separated in this way? Although these are not separated, it is "necessary that thus the whole mass be moved" into the place left by those fluid parts that are being drawn outside of the body "so that all places may be filled" (OL I.366; EW I.450). When the non-fluid parts of the body endeavour this way, the body swells: "[. . .] the mass of the body, all striving at the same time in that way, swells" (OL I.366; EW I.450). Hobbes has now arrived at a "possible cause" of the felt heat of the sun. When humans become warm and begin to sweat after sitting in the sun, they do so neither because the air around them is hot nor because the sun itself is hot. Instead, warming and sweating occurs because the parts of the air around them are continually changing place (fermenting), causing the liquid parts of human bodies to leave and other parts to swell.

This may seem like a strange explanation; for it might appear that these fluids simply exit the body because *qua* fluids they do so more easily than other parts of the body. However, the reason *why* these fluids exit the body is found in the account of fermentation in *De corpore* XXI.5 (OL I.263–265; EW I.323–325). These fluids exit the body because they are being separated from the non-fluid parts of the body and, through fermentation, are being joined with other fluids. A consequence of fermentation's seething is that homogeneous fluid bodies are united, a property of this type of motion that speaker A in *Examinatio* later highlights (OL IV.228). The fermentation process, whether in the case of human sweat, cloud formation, or, as Hobbes also mentions, young wine, need not be caused by fire (OL I.264; EW I.324–325). Such heat is produced because of the circular motion involved in the perpetual change of place.

Hobbes uses 'possible' to signal that the sun may not *actually* cause felt heat by means of simple circular motion. The same level of certainty in the current explanation was accorded to the explanation of sense: we know that *if* the sun moves the air around it by simple circular motion, causing fermentation, *then* the fermentation motion continues to the human body and causes felt heat in the sentient body when the fluid parts leave and the non-fluid parts swell to prevent vacant spaces.

Thus, the supposition that the sun moves the ether around it by simple circular motion makes this a possible cause.[24]

These two natural-philosophical explanations from *De corpore* Part IV described so far—the explanations of sense and heat and swelling—display Hobbes's mixing of experience (the 'that') and borrowed maker's knowledge from geometry (the 'why'). When borrowing a geometrical principle in natural philosophy, we must *suppose* that bodies move according to that principle. This borrowing from geometry would be difficult to explain on the deductivist interpretation of the relation between Hobbes's geometry and natural philosophy, for in neither explanation do we find a deduction nor the suggestion that one could be provided from geometry to these explanations. Instead, I have argued that the borrowing in these explanations, and elsewhere in *De corpore* Part IV (see fn. 4), should be understood as part of mixed-mathematical explanations, making them examples of the "true physics" that Hobbes describes in *De homine* X.5.

We find additional support for the view that "true physics" is mixed mathematics in Hobbes's comments about failed or limited explanations. In some explanations, Hobbes thinks that since the 'that' is insufficiently known it is useless to seek the 'why'. For example, even though Kepler posits a cause for the eccentricity of the Earth's orbit, Hobbes holds that the 'that' is insufficiently known: "But since the *hoti* is not yet evident, it is in vain for the *dioti* to be searched for" (Hobbes 1655, 254).[25] Although Hobbes believed it was first necessary to have the *a priori* principles of first philosophy and the causal principles of geometry in place, when seeking to provide an explanation in natural philosophy he also saw the need for making sure the 'that' was sufficiently well known before trying to posit a possible cause at all. Hobbes's comments on the relationship between the 'that' and the 'why' help make sense of his claim that natural history is "[...] very useful (no, indeed necessary) for philosophy [...]" (OL I.9; EW I.10; Hobbes 1981, 189; see also *De homine* XI.10, OL II.100) while also excluding natural history from philosophy proper.

5.3 Explaining the Behavior of the Air-Pump in *Dialogus Physicus*

Hobbes's debate with Robert Boyle places the status of knowledge to be gained by experience/experiment into focus. This section examines Hobbes's view of the

---

[24] Hobbes elsewhere identifies the simple motion of the sun as a "supposition" (OL I.351; EW I.431; see Horstmann 2001, 494–495).

[25] This claim is present in the first edition of *De corpore* (1655) and transmitted to the English edition of *Concerning Body* (1656b, 329) and to the *English Works* edition (EW I.443). However, the *Latin Works* edition (OL I.361) does not contain it, following the 2nd Latin edition of *De corpore* (1668); Schuhmann follows the OL (Hobbes 1999, 301).

152   HOBBES'S TWO SCIENCES

proper method for discovering possible causes by looking to Hobbes's criticisms of the conclusions that Boyle attempted to derive from the air-pump experiments. I argue that Hobbes's treatment of the nature of air in the *Dialogus Physicus* reflects his understanding of natural philosophy as mixed mathematics that has been articulated in the previous sections of this chapter. Given this account of natural-philosophical explanation, Hobbes held that air-pump experiments could establish only the 'that'; any possible cause must be constrained by the principles of first philosophy, and the reason 'why' must be borrowed from geometry, which Hobbes does when he cites principles related to simple circular motion in *Dialogus Physicus*. Hobbes identifies principles of motion such as this as "easy and *mathematical*" within the text of *Dialogus Physicus* (Hobbes 1985, 273; emphasis added).[26] In the subsections that follow, I show Hobbes's borrowing of causal principles from *De corpore* and contrast the mixed-mathematics understanding with Shapin and Schaffer's view that Boyle sought to defend against the Hobbesian "beast of deductivism" (Shapin and Schaffer 1985, 176).

5.3.1 Suppositions and Simple Circular Motion

Hobbes published *Dialogus Physicus* (1661) at a key point in his life. He finished *Leviathan* in 1651 and was thus able to return to pursuits in natural philosophy, leading to the publication of *De corpore* in 1655. He had also by this point published *De Homine* (1658), the middle volume of the *Elementa* trilogy that he published last. Given its place within Hobbes's active writing period, there are explicit citations to *De corpore* within the discussion of the *Dialogus Physicus*. In citing himself in this work, I suggest that Hobbes is not merely repeating himself to emphasize that he has *already* proposed a solution to some problem elsewhere in his voluminous writings, something he sometimes does in contexts such as *Examinatio* (1660; OL IV.84).[27] Instead, I argue that these citations of principles from *De corpore* in *Dialogus Physicus* are examples of Hobbes borrowing geometrical principles—the 'why'—within a mixed-mathematical explanation.

The *Dialogus Physicus* begins with a dedication of the work to Samuel Sorbière. In his typically unabashed manner, Hobbes proclaims that those who are making experiments at Gresham College "may meet and confer in study and make as many experiments as they like, yet unless they use my principles they will advance nothing" (Hobbes 1985, 347). But in addition to Hobbes's irascibility and hubris, behind the claim that they will "advance nothing" was his view that natural

---

[26] Citations to *Dialogus Physicus* will be to Simon Schaffer's translation of the 1661 edition of *Dialogus Physicus* in Shapin and Schaffer (1985). This work is also reproduced in the *Latin Works* edition (OL IV).
[27] See Jesseph (2004, 202–203) for discussion of this example from *Examinatio*.

philosophy must be done according to the proper method: "[. . .] ingenuity is one thing and method [*ars*] is another." Proper method, Hobbes argues, requires that one "investigate the causes of those things done by motion through a knowledge of motion, the knowledge of which, the noblest part of geometry, is hitherto untouched." Hobbes sees himself as having given just this sort of knowledge of motion considered as part of geometry, but since the experimenters of Gresham College have, in Hobbes's mind, neglected his work, he states that his work has been for naught: "[. . .] as yet it seems I live in vain" (Hobbes 1985, 347). In another dialogue—*Examinatio*—one of the speakers complains that those who are devoted to philosophy have come to "focus solely on acquiring new phenomena" and have "neglected the contemplation of the nature of motion" (OL IV.228).

The dialogue in *Dialogus Physicus* occurs between two participants—A and B— and participant B is supposed to represent a member of the meetings at Gresham College. After discussing the construction of the air-pump, interlocutor B questions A regarding what he thinks follows from the experiments, asking "Would not the space left by the sucker be a vacuum?" (Hobbes 1985, 353). Speaker A replies that the answer to this question requires that we first know the nature of air.

Speaker A, as Hobbes's mouthpiece, supposes that air is a liquid that is capable of infinite divisibility: "[. . .] if a part of the air, whose quantity is less than any water-drop you have seen, is fluid, how is it to be proved to you by anyone that a part half the size of its parts [. . .] might not be of the same nature?" (Hobbes 1985, 353–354; OL IV.244). The claim holds, A asserts, even for parts of air "one hundred thousand thousandeth." B concedes A's claim that the nature of air must be supposed before a judgment can be made regarding what remains in the space left by the sucker; he notes that he and the others engaging in the experiments distinguish fluids from nonfluids by the "size of the parts of which any body consists" (Hobbes 1985, 354; OL IV.244). With this view, ashes and dust count as liquid and, furthermore, B holds that liquids may be composed of nonfluids. Thus, upon division it might be discovered that a fluid such as air may be composed from non-fluid parts.

When B asserts that the experimenters do not "stomach [. . .] infinite divisibility," A grants that infinite division (*divisio*) itself cannot be conceived but that infinite divisibility (*divisibilitas*) can easily be conceived (Hobbes 1985, 354; OL IV.244– 245). To support this claim regarding the conceivability of infinite divisi*bility*, A appeals to God's abilities and asserts that it is no more difficult for God to "create a fluid body less than any given atom whose parts might actually flow" than it is for God to create the ocean (Hobbes 1985, 354; OL IV.244–245). So God could create water such that all of its parts, no matter how small, are also water. This reference to "almighty God" as grounding that which shows whether a claim about something's nature is conceivable is odd given Hobbes's view on our limited knowledge of God as first cause, but I will leave this to the side.

If we grant this point to Hobbes's apparent spokesperson, speaker A, as B does, then we should agree as well that "it is not necessary for the place that is left by the

154   HOBBES'S TWO SCIENCES

pulling back of the sucker to be empty" (Hobbes 1985, 354; OL IV.245). Given the conceivability of "infinitely subtle" parts of air, we cannot be certain that air does not enter between small gaps between the sucker and the cylinder, thus calling into question the claim that "the space left by the sucker" must be taken to be a vacuum (Hobbes 1985, 353).

B's rebuttal to A regarding this possibility of infinitely subtle bits of air entering by means of small gaps appeals directly to experience: "But when the tap was turned, we observed that a sound was made as if air were breaking into the cylinder" (Hobbes 1985, 355; OL IV.246). The sound, B claims, counterweighs against A's proposal, but A has a reply at the ready: the sound heard was caused by the collision of the air in the cylinder with the air outside of it. In the remainder of this subsection, I will examine two suppositions that arise in the resulting disagreement: the supposition held by B that air possesses an elastic force that accounts for its behavior in appearing to resist compression, and the supposition held by A, which relies upon borrowed principles from *De corpore*, that air moves according to simple circular motion.

Interlocutor B explains the sound made when the tap is turned by supposing that there is an "elastic force" in the air; air is such that its parts are "endowed with this nature" (Hobbes 1985, 355; OL IV.247). To understand this, B asks A to imagine that the air nearest to the surface of the Earth is "like a heap of corpuscles" that behave like wool fibers. The spring in air from its elastic force causes it to bounce back after compression, similar to the way that wool bounces back after being pressed between one's hands. We can understand both behaviors by supposing the parts of both to be "endowed with a power or principle of dilation" (Hobbes 1985, 356; OL IV.247). After A and B agree that "all things supposed" in hypotheses must be conceivable, A questions B's hypothesis about the supposed principle of dilation, asking how experimental philosophers like B could have gained knowledge of such a cause from merely observing phenomena like crossbows and wool.

B admits that the cause of elastic force being supposed is not "very certain." Nevertheless, it is certain that the bodies in question—wool and crossbows— cannot *self*-move (B is thus endorsing as a constraint the persistence principle from *De corpore* VIII.19 discussed in subsection 5.1.2). Furthermore, B asserts that the motion cannot be straight motion, for if it were, "the whole body (so to speak) would be carried away by the motion of the crossbow itself [. . .] [t]herefore it is necessary that the endeavour be circular, such that every point in a body restoring itself may perform a circle" (Hobbes 1985, 357; OL IV.248–249). Hobbes, speaking as A, disagrees with B and claims that this assertion about circularity is "not necessary." Instead, A asserts that it must be a motion that returns the body to its starting point. Granting A's point that the motion must be one of restitution—this is the motion that will eventually explain the sound when the tap of the air-pump was turned—B posits that perhaps the source of this restitutional motion must be located in particulars smaller than air, which smaller parts have "their own natural motion of which there is no beginning" (Hobbes 1985, 358; OL IV.249).

Speaker A proposes that the possible cause of this restitutional motion by the parts of crossbows and wool fibers relates to a type of motion that Hobbes develops as part of his geometry in Part III of *De corpore* in chapters XXI.1 and XXI.10 (more on simple circular motion as part of geometry below).[28] Hobbes, as A, further claims that the only way to hold that there is an "elastic force" in air is to hold this supposition; without supposing the Hobbesian possible cause, it is impossible. Given the earlier remarks regarding "conceivability" as a constraint for suppositions, Hobbes means that without the supposition of circular motion by smaller parts of bodies the supposition related to the "elastic force" of air is impossible because it is inconceivable, that is, unimaginable. The project in the next section of the dialogue is to argue, on the one hand, that this "natural motion of which there is no beginning," which A will argue is simple circular motion, is conceivable while, on the other hand, arguing that elastic force is not conceivable.

A's introduction of simple circular motion into the dialogue relates to two distinct parts of the Hobbes's project in *De corpore* discussed already in this chapter: first, Hobbes's understanding motion as part of pure geometry (Part III of *De corpore*); and second, his use of geometrical principles within natural philosophy as suppositions for how bodies *may* behave (Part IV of *De corpore*). Distinguishing the different roles that discussions of 'motion' play in *De corpore* makes A's very brief comments about simple circular motion in the *Dialogus Physicus* clearer. The first part of the *De corpore* project to which speaker A refers are the (pure) geometrical consequences of simple circular motion that Hobbes demonstrates in *De corpore* XXI, of which speaker A explicitly cites articles 1 and 10 (see fn. 28). First speaker A cites a principle about how any body behaves when moving in simple circular motion: when a body moves in simple circular motion, all its points describe the circle the body makes (from *De corpore* XXI.1). Next, A refers to Hobbes's claim that simple circular motion is generated from simple circular motion (from *De corpore* XXI.10). These two citations are from Part III of *De corpore* on geometry, but within the same paragraph, interlocutor A discusses how these geometrical principles are *used* within another section of *De corpore*, Part IV

---

[28] A textual note is in order. There appears to be an error in the printing and translation of *Dialogus Physicus* regarding this reference to *De corpore* XXI. Both Molesworth (OL IV.251) and Schaffer's translation (Hobbes, *Dialogus Physicus*, 360) reproduce these references as being to *De corpore* II.1 and II.10 from the original printing (cf. Hobbes 1661, 9), but this cannot be the passage to which Hobbes, as interlocutor A, intended to refer since Hobbes does not discuss motion at all in *De corpore* II.1 or II.10. Instead, in those articles he discusses names. It is possible that the 1661 edition of *Dialogus Physicus* introduced this error in printing since there is an odd space between the '2' and the decimal point that is not present in the other references to *De corpore* within that edition of the text (cf. Hobbes 1661, 9, which reads "[...] (ut ab eo demonstratum est *Lib. deCorp.Cap.2 .Artic.*I.) [...]"). Thus, the references in this section of *Dialogus Physicus* (Hobbes 1985, 360; OL IV.251) clearly refer to *De corpore* XXI.1 and XXI.10.

156   HOBBES'S TWO SCIENCES

on natural philosophy, where they become suppositions about how actual bodies (conceivably) may behave.

Speaker A gestures to Hobbes's usage of these geometrical principles related to simple circular motion by mentioning how Hobbes connects them to the Copernican hypothesis in *De corpore* Part IV.[29] A notes that Hobbes posits two simple circular motions to explain the diurnal and annual motions of the Earth. Hobbes explains the Earth's annual motion by supposing that the sun has in its nature a simple circular motion and that it moves the earth around it with this motion. Given the supposition that the sun moves in this way, Hobbes can claim that it moves the Earth circularly around it because, as discussed above, simple circular motion is generated from simple circular motion (from *De corpore* XXI.10); this claim about the motions of the Earth relies upon the inference made in the geometry of Part III. Similarly, Hobbes supposes that the Earth itself has its own simple circular motion, which A says is "due to its own nature or creation" (Hobbes 1985, 360; OL IV.252), allowing him to explain the Earth's diurnal motion. Speaker A will use this second supposition relating to the Earth's natural circular motion to explain the nature of air.

A next asks B what would result if the Earth were "annihilated by divine omnipotence or if half this Earth were removed to some other distant place beyond the fixed stars" (Hobbes 1985, 361; OL IV.253). B agrees with A that any part of Earth—in B's words even "one of its atoms"—would retain the simple circular motion that is "congenital" to the Earth. In other words, if simple circular motion is congenital to the Earth considered as a whole, any of its parts would retain this same circular motion. Thus, if we take the air as having "particles of earth and water [. . .] interspersed" throughout it, we can then use these hypotheses to explain the sound made when the tap of the air-pump is turned. B grants that the "hypotheses are by no means absurd," but B requests that A further show how they help in saving two phenomena related to the air-pump, which B still takes to be better explained by positing an elastic force in the air.

The first *explanandum* relates to B's observation that he has "seen the handle fall back from the hands of whoever happens to pull back from the sucker" after the suction has been engaged repeatedly (Hobbes 1985, 362; OL IV.254). A's treatment of the phenomenon straightforwardly follows from understanding air as a combination of particles of earth and water. He claims that the continual motion of the sucker caused pure air to be drawn into the cylinder while not allowing earthy particles to be drawn in. Drawing pure air into the cylinder caused a "greater ratio of earthy particles that were near the sucker outside the cylinder

---

[29] Schaffer (Hobbes 1985, 360 fn. 22) notes Hobbes's supposition that simple circular motion is responsible for the motions of the Earth in *De corpore* XXVI.5–7 (Part IV), but this misses that A's introduction of the principles related to simple circular motion draws upon the Hobbes's section on geometry (Part III).

to the pure air in which they exercised their motion than before" the sucker had been engaged (Hobbes 1985, 362; OL IV.254). Since the ratio is greater following this suction, the particles closest the sucker press upon it, which A claims "is the phenomenon itself."

Since both A's and B's hypotheses save the phenomenon of the handle of the sucker being pulled away from the hand of the user, A explains why his—Hobbes's— is to be preferred. To make this choice, A emphasizes two constraints on hypothesis choice: first, that the hypothesis be "conceivable"; and second, that if one assumes a hypothesis, then the effect must follow necessarily. The first criterion, A claims, is the deciding factor between his hypothesis and B's: "[y]our hypothesis lacks the first of these; unless perhaps we concede what is not to be conceded, that something can be moved by itself" (Hobbes 1985, 362–363; OL IV.254–255). As Hobbes understands 'inconceivable', this implies that B's hypothesis that air self-moves by an elastic force is unimaginable. Here Hobbes, through A, is implicitly relying upon the persistence principle from *De corpore* VIII.19 (discussed in subsection 5.1.2) as a constraint upon what can function as a possible cause in natural philosophy. In contrast, in Hobbes's understanding of 'conceivable', we can easily imagine that with less space to "exercise their natural motion" the earthy particles would naturally press against the sucker.

The second *explanandum* B highlights is that after the sucker is repeatedly operated, it is difficult for anyone to remove the cover of the upper orifice; it is "as if a weight of many pounds hung from it" (Hobbes 1985, 365; OL IV.257). To explain this, A appeals to a "very strong circular endeavour of the air, made by the violent entry of the air in between the convex surface of the sucker and the concave surface of the cylinder." The repeated action of the sucker stirs up the air into a more violent circular motion than usual, and this more violent motion of the air in the cylinder makes it more difficult to remove the cover. If the "air were at rest," removing the cover would not be so difficult. B concedes that this is a plausible explanation and both interlocutors agree that the swiftness of the air inside the cylinder could also explain why bubbles appear inside the cylinder when the sucker is engaged, apart from any addition of heat (Hobbes 1985, 365; OL IV.258). Similarly, A argues that the behavior of wind-guns can be explained by the violent circular motion of air impacted by the force of a sucker (Hobbes 1985, 368; OL IV.260); the violently-moving air pushes out of the chamber forcefully when released and shoots the ball.

The interlocutors discuss several other phenomena in the remaining discussion, but I shall only mention one additional phenomenon that is most relevant to the discussion in the next subsection. After B admits that A has provided a plausible explanation of the behavior of both the wind-gun and the air-pump, B questions A on how his understanding of the nature of air in terms of circular motion can explain what the experimenters have witnessed concerning the weight of bladders. B reports that the experimenters have hung an inflated bladder from the cover of the

158   HOBBES'S TWO SCIENCES

air-pump and placed a scale inside the device. After engaging the sucker, they observe that the bladder appears to weigh more, and they attribute this to the air inside the bladder weighing more when weighed in an "empty space" (Hobbes 1985, 368; OL IV.261).

Notably, Hobbes as speaker A does not question the experimenters' report. He notes: "They can be certain that the scale in which the bladder is, is more depressed than the other, their eyes bearing witness" (Hobbes 1985, 369; OL IV.261). However, A argues that they cannot know that the gravity of air is the *cause* of what they observe. A argues that it is plausible that a "greater quantity of atoms from the bellows or sooty corpuscles from the breath being blown in" could be responsible for the greater weight. To explain the observations related to the greater weight of the bladder following the action of the sucker, A draws upon Hobbes's account of fermentation from *De corpore*. As discussed already, Hobbes thinks of fermentation as a type of "seething" circular motion whereby heterogeneous particles are separated from one another and homogeneous particles are joined together (more on fermentation below). Hobbes develops this geometrical principle of fermentation in *De corpore* XXI.5 (Part III of *De corpore*), which A cites here in *Dialogus Physicus*, and then he posits that the sun moves the air around it in simple circular motion and causes fermentation. With this geometrical principle and the added supposition that the sun moves in this way, Hobbes explains various phenomena in *De corpore* Part IV, such as his explanation in *De corpore* XXVII.3 for why human bodies sweat and swell when warm (discussed in subsection 5.2.2).

In similar fashion, speaker A explains the greater weight of the bladder by suggesting that we understand the increase in weight as due to fermentation moving homogeneous particles together. Attributing the greater weight to the "cause of gravity" is insufficient, A contends, because it "could not bring together the homogenous substance when separated by force and tear apart heterogeneous bodies brought together by force" (Hobbes 1985, 369; OL IV.262). Interlocutor A's point is that simply positing that air has weight because of its "natural gravity" will not account for why the bellows full of air weighs *more when the sucker is engaged*. In other words, gravity could not explain why there were more particles of air in the bladder after the action of the sucker, but Hobbes's principle of fermentation could if we supposed that the sucker action caused a fermenting motion of the air. This fermenting motion of the air would then join homogeneous particles of air, one with another, causing the bladder to be filled with a greater number of particles of air than there were prior to the action of the sucker. This is not the place to offer criticisms of the explanation, but Hobbes is right, of course, that *simply* appealing to a "natural gravity" would not explain the increase in weight; however, his proposal is not without problems of its own. For example, given A's supposition that the sucker causes the air to move by fermentation, how would interlocutor A explain why the homogeneous particles of air go *into* the bladder rather than *out of* the bladder to be joined with like particles outside of the bladder?

This subsection has focused upon two features of the *Dialogus Physicus*. First, it has examined the detailed description of the experiments and apparatuses of the air-pump experiments. Importantly, Hobbes as interlocutor A does not deny the veracity of the experimenters' reports but instead questions the explanation they provide by focusing on the causal suppositions to which they appeal. Second, this section has drawn attention to interlocutor A's explicit citations of geometrical principles from Part III of *De corpore* within the *Dialogus Physicus*, which occur in just the same way as the borrowing of geometrical principles within the natural philosophy of Part IV of *De corpore*. A two-part approach to natural philosophy is thus evident in both *De corpore* and *Dialogus Physicus*: first one establishes on the basis of experience/experiment *that* a phenomenon occurs, and then one explains *why* by borrowing a principle from geometry. In the next subsection, I argue that Hobbes's conception of geometry as *scientia* and natural philosophy as mixed mathematics makes sense of interlocutor A's, as Hobbes's mouthpiece, explanation of the phenomena surrounding the air-pump experiments as well as his criticisms of others' conclusions made from them. In the final subsection, I reassess Hobbes's putative "deductivism" in light of the debate concerning the air-pump experiments and his natural philosophy more generally. Readers interested primarily in the main argumentative points related to the *use* of mathematics within natural philosophy may skip subsection 5.3.2 since it deals with Hobbes's "easy" and "mathematical" demonstration of some of the properties of "simple circular motion," that is, it shows why Hobbes believed he had given a mathematical treatment of that type of motion.

5.3.2 Mixing Experience with the "Easy" and "Mathematical" Doctrine of Motion

After discussing these various phenomena witnessed by the experimenters, interlocutor A pinpoints what he takes to be the hopelessness of the attempts by Boyle and others at advancing physics: "[. . .] how did you dare take such a burden upon yourselves, and to arouse in very learned men [. . .] the expectation of advancing physics, when you have not yet established the doctrine of universal and abstract motion (which was easy and mathematical [*quod facile et mathematicum erat*])?" (Hobbes 1985, 379; OL IV.273). This description of principles of motion, such as those related to simple circular motion, as *mathematical* is unsurprising given the account of Hobbesian natural philosophy as mixed outlined so far.

Nevertheless, it may not be immediately obvious what Hobbes means when he says that "the doctrine of universal and abstract motion" is both "easy and mathematical." As a part of mathematics, Hobbes views the doctrine of motion as "easy" because we are able to gain *actual* causal knowledge when we make geometrical constructions related to motion (as discussed in chapter 3), something which is

160   HOBBES'S TWO SCIENCES

unavailable in natural philosophy where we have only knowledge of *possible* causes. Hobbes illustrates this point about the special status of "easy" knowledge gained from construction in mathematics in *De corpore* I.5. He argues that examining a circle by "sense" cannot give one knowledge of that figure's properties—one cannot know whether a figure observed by sense is actually a circle (OL I.5; EW I.6). One could know that it is circular insofar as it *appears* to be a circle, but one could not know whether, in fact, it has all the properties of a circle because one did not construct that circle. However, the person who has *constructed* this circle is in an epistemically privileged situation, for this individual sees the causes as she constructs the figure. Determining all of the properties "from the known generation of the displayed figure, is *most easy* [*ex cognita figurae propositae generatione, facillime*]" (OL I.5; EW I.6; emphasis added). Since Hobbes assumes that knowing the causes of a thing entails knowing all of its properties (Gauthier 1997, 512), once one has constructed a figure, the task of knowing that figure's properties is "most easy." In *Six Lessons*, Hobbes suggests that if someone had never seen a circle constructed they might not even believe such a figure was possible (EW VII.205).

Hobbes's use of 'easy' in *Dialogus Physicus* should be understood in this light. The causal principles related to simple circular motion that interlocutor A uses to explain the behavior of air are easy since they are derived from constructions related to motion. Even if "easy," why would Hobbes hold that the "doctrine of motion," and in particular simple circular motion, is part of *mathematics*? Hobbesian definitions of geometrical figures must instruct how to make those figures—they must incorporate *motion* within the definitions themselves. Hobbes's already discussed definition for 'line' can serve as an example: "a line is made from the motion of a point" (OL I.63; EW I.70; Hobbes 1985, 297).

However, notice that these definitions of simple geometrical figures, such as the definition of 'line', employ the concept of motion only as *locomotion*. Beyond his claim that all geometrical definitions must include motion, Hobbes, like Barrow, views motion *itself* as part of geometry. Since motion and magnitude are the "most common accidents of all bodies" (OL I.75; EW I.203), he unsurprisingly titles *De corpore* Part III "Proportions of Motions and Magnitudes." Beyond simple locomotion, Part III first defines different types of motion in terms of the kinds of bodies that move (*De corpore* XV.4); and second, develops the principles of various kinds of motion by means of constructed geometrical diagrams. In what remains of this subsection, I examine these two aspects of Part III and, with regard to the second aspect, focus upon Hobbes's development, by means of construction, of simple circular motion in *De corpore* XXI.

De corpore Part III begins with a review of the principles established in the preceding section and introduces additional principles, including those related to endeavor and impetus. Hobbes justifies including 'endeavor' within geometry by describing how one may compare one endeavor with another, just like one may compare points of inequality for lines that intersect concentric circles:

[...] if a straight line cut many circumferences of concentric circles, the inequality of the points of intersection will be in the same proportion which the perimeters have to one another. And in the same manner, if two motions begin and end both together, their endeavours will be equal or unequal, according to the proportion of their velocities; as we see a bullet of lead descend with greater endeavour than a ball of wool. (OL I.78; EW I.206–207)

Hobbes develops 'simple motion' by noting that we "consider" bodies in various ways. As argued in chapter 3, Hobbes consistently uses 'consider' to describe the status of mathematical objects as relying upon our interests and goals. Here Hobbes distinguishes *simple* and *compound* motion by noting that we may "consider" bodies as either having parts or not, depending upon our interests. He claims that "[...] when a mobile is considered [*consideratur*] as having parts, there arises another distinction of motion into *simple* and *compound*. *Simple*, when all the several parts describe equal lines; *compounded*, when unequal" (OL I.181; EW I.215). Whether 'simple' or 'compound' applies to motions depends upon how we consider a body.

De corpore XXI.1 begins by reiterating this definition of simple motion and adding to it with consideration of simple *circular* motion: "[...] in simple circular motion it is necessary that every straight line taken in the moved body be always carried parallel to itself [...]" (OL I.259; EW I.318). By "simple circular motion" Hobbes means something like the motion of a sieve, a sort of gyrating motion around a center point.[30] Hobbes provides several analogies to help the reader understand simple circular motion. He compares simple circular motion in *De corpore* XXI.2 to what would happen "[...] if a man had a ruler, in which many pens' points of equal length were fastened, he might with this one motion write many lines at once" (OL I.261; EW I.320). For contemporary readers, a chalk holder that a music teacher uses to draw a five-line music staff on a chalkboard is an example similar to what Hobbes has in mind. As mentioned already, in *Examinatio* he provides an analogy to a ball with a stylus attached to it (OL IV.227) and considers the motion of the ball while someone is writing with continuous motion. Hobbes demonstrates the claim that "every straight line taken in the moved body be always carried parallel to itself" with a geometrical construction with figure 1 in the 1655 printed text.

In the articles that follow this definition of simple circular motion (*De corpore* XXI, articles 2–6), Hobbes uses another constructed diagram (see Figure 5.2). Using this diagram, he first demonstrates that in circular motion (not *simple*

---

[30] Some have argued that Galileo was Hobbes's source for this understanding of simple circular motion (see fn. 19; more on this below). In the demonstration to be discussed below from *De corpore* XXI.2, Hobbes does not appeal to the motion of C–D–E as a gyrational simple circular motion, but only as a contrary revolution.

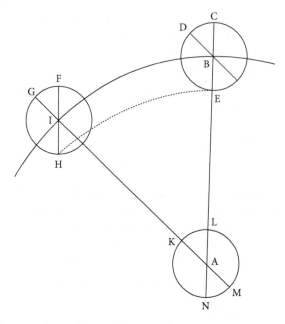

Figure 5.2 The Axis of a Circle Moved by Circular Motion.

circular motion) the axis of an epicycle, such as the axis E–C of epicycle C–D–E, will always be carried parallel to itself when moved around a center point such as A and when revolving around its own center point such as B.

Hobbes demonstrates this claim using the figure by asking the reader to suppose that equal angles are made in equal time. On that assumption, as the angle B–A–I is made when C–D–E moves leftward in the diagram in revolution around K–L–M–N, it will require the same amount of time to make any angle equal to it, such as angle F–I–G. Were there no contrary revolutionary motion of C–D–E, which causes point E to move toward C through D, the radius B–C would end at G–I as B–I is traversed. However, the contrary revolutionary motion causes C–B to end at I–F. Here is Hobbes's description: "Wherefore, in what time B C comes to I G by the motion from B to I upon the centre A, in the same time G will come to F by the contrary motion of the epicycle, that is, it will be turned backwards to F, and I G will lie in I F" (OL I.260; EW I.320).

Hobbes draws a corollary from this demonstration. He suggests that his account of simple circular motion will allow him to save the phenomena without needing the two annual motions of the Earth proposed by Copernicus. In other words, he thinks that he needs only "one circular simple motion" to account for the Earth's axis remaining parallel to itself without needing a third motion beyond the annual movement around the sun and diurnal rotation. This immediate application of the principles of simple circular motion to phenomena such as the Earth's motion might make it strange to claim that the demonstration proceeding from that application is *geometrical*. Were Hobbes being more circumspect, he would not have made this a corollary and instead consistently talked only of circles, i.e., bodies *considered* as circles, and made claims only that apply to "any solid body [*quodlibet corpore solido*]" (OL I.259; EW I.318), as he does with figure 1 in *De corpore* XXI.1. I suggest that we take this slide into application, or more precisely "mixing," to be in the service of making the geometrical section more accessible to readers. Indeed, this way of understanding the example makes sense of the fact that immediately following the brief discussion of how simple circular motion simplifies the Copernican system, Hobbes connects it to sieve-like motion, i.e., "the same [motion] which is used by all men when they turn anything round with their arms, as they do in grinding and sifting" (OL I.261; EW I.320).

After this example and its corollary, Hobbes seeks to "set down some properties" of simple circular motion (OL I. 261; EW I.321). Hobbes uses the diagram above (Figure 5.2; figure 2 from the 1655 edition) to demonstrate the first three properties, all of which relate to the properties of bodies moving according to simple circular motion in fluid media. The first property, which he demonstrates in *De corpore* XXI.3, is as follows: "[. . .] there are no parts so small of the fluid ambient, how soever far it be continued, but do change their situation in such manner, as that they leave their places continually to other small parts that come into the same" (OL I. 261; EW I.321).

He first asks the reader to consider a body, such as K–L–M–N, being moved with simple circular motion in a fluid medium. Given his understanding of simple circular motion as a sieve-like motion wherein all points of a body move around a center point, it follows that all points of a body so moved create lines parallel to each other. As K–L–M–N moves in this way, Hobbes argues that "the centre A and every other point [in K–L–M–N], and consequently the moved body itself, will be carried sometimes toward the side where is K, and sometimes toward the other side where is M" (OL I.261; EW I.321). Assuming a plenum, Hobbes argues that when K–L–M–N is so moved all parts of the fluid medium will move as well, no matter how small, so that their place can be taken by K–L–M–N.

The second property that Hobbes demonstrates about simple circular motion in *De corpore* XXI.4 is that when all the parts of a fluid medium are moved with simple circular motion by a body moving so in the medium, we should think of their motions in terms of concentric circles radiating out from the center. The parts farther from the center will make their circles in longer times than those closer, and we can understand the relationships among these times as being in "[. . .] the same proportion with their distances from the movent" (OL I.263; EW I.322).

It is the third property, fermentation, which has been discussed in subsection 5.2.2 related to the explanation of sweating and swelling in *De corpore* XXVII, that relates most directly to the discussion of the nature of air in *Dialogus Physicus*. Hobbes asks the reader to consider a body with the "same simple motion" already

164   HOBBES'S TWO SCIENCES

considered for body K–L–M–N. Hobbes reminds the reader that bodies are not different from one another insofar as they are bodies but only insofar as they differ in their respective motions, from "some internal motion, or motions of their smallest parts" (OL I.264; EW I.323). This view that both likeness and difference of one body to another is explained exclusively in terms of a difference or similarity in internal motions is a view that follows straightforwardly from a principle in *De corpore* IX.9, which Hobbes cites. As a body is moved with simple circular motion in a fluid medium, Hobbes claims that this continual motion puts together homogeneous bodies, understood as those bodies with similar internal motions.

Likewise, simple circular motion drives apart heterogeneous bodies, those with dissimilar internal motions. Hobbes says that this property of simple circular motion—that it results in the joining of homogeneous bodies and driving apart of heterogeneous ones—is the cause of what is commonly called "fermentation." He traces the etymology of the term to *fervere* and relates it to the meaning of "seething." Hobbes takes it that the cause of this seething in fermentation is due to "all parts of the water chang[ing] their places" so that as a result "[...] the parts of any thing, that is thrown into it, will go several ways according to their several natures" (OL I.264; EW I.324).

What makes these demonstrations related to simple circular motion *mathematical*? It would be uninformative to say that they are mathematical simply because they are contained within Part III of *De corpore*, which Hobbes says concerns geometry. Furthermore, they are not geometrical merely because they concern demonstrations about bodies considered abstractly, i.e., bodies considered apart from their unique qualities like "shining" (see discussion of *Anti-White* I.1 above; Hobbes 1973, 106; Hobbes 1976, 244–225). In addition to being about bodies considered abstractly, these are geometrical because *we make the figures*—we manipulate the figures in the construction, such as circle C–D–E, and as we move them from center B to center I we gain causal knowledge. Furthermore, each step of the demonstrations relies only upon considering the two features that Hobbes describes as characterizing geometry because they are the most common accidents of all bodies, *motion* and *magnitude*. In other words, they are abstractly-considered bodies at the right level—the subject of geometry at the level of magnitude and motion—and not merely an abstraction. Indeed, for Hobbes, merely considering bodies abstractly as having magnitude without motion would not lead one to geometrical principles. Furthermore, attempting to make a claim about motion *without* extension would cause one to make absurd claims, in Hobbes's terms. Thus, I suggest that Hobbes understands these principles relating to simple circular motion as geometrical because they are about abstractly-considered bodies, those with magnitude and in motion, and because we arrive at them by engaging in the making of diagrams.

Thus, Hobbes can say that makers possess causal knowledge (*scientia*) of the principles related to simple circular motion. In the *Dialogus Physicus*, interlocutor

A argues that attributing the greater weight of the bladder to the "cause of gravity" is insufficient because it "could not bring together the homogenous substance when separated by force and tear apart heterogeneous bodies brought together by force" (Hobbes 1985, 369; OL IV.262). If instead we posit that the sucker causes fermentation, then we may borrow the geometrical principle from Part III of *De corpore*. Borrowing this causal principle will allow us to explain *why* the weight increases— it is due to the motion of fermentation joining together homogeneous particles.

Such a borrowing lends a higher degree of plausibility, Hobbes would contend, to A's explanation. Since fermentation motion is a consequence of continual simple circular motion, we can know that if a body (like the sucker of the air-pump) causes the simple circular motion of a medium through continuous motion (repeated movements of the sucker), then fermentation will result. We cannot know with certainty that fermentation is in fact responsible (since we are not the makers of natural things); nevertheless, we can know that *if* continuous simple circular motion were occurring, then fermentation would necessarily occur and also that *if* fermentation were involved, then the joining of homogeneous particles would necessarily occur. As I suggested above for the two explanations from *De corpore*, the result is that a natural philosopher may have suppositional certainty for their explanations.

This is Hobbes's *method* that he emphasizes cannot be replaced with mere *ingenuity* on the part of Boyle and colleagues, whose fascination with artificial experimental phenomena Hobbes compares in *Examinatio* to a child who "contemplates the beauty of a book more than its letters" (OL IV.228). Experiments do "enrich natural history," Hobbes grants, and without them "the science [*scientia*] of nature is pursued in vain"; however, if experiments (*experimenta*) are sufficient for scientific knowledge, then "pharmacists are the best physicists of all" (OL IV.228). Hobbesian natural philosophy must begin with consideration of bodies as magnitude in motion, demonstrate necessary consequences of those bodies in motion (like fermentation), and finally borrow those consequences as causal principles within explanations of natural phenomena.

5.3.3 Deduction, Deductivism, and the *Dialogus Physicus*

Shapin and Schaffer's influential work *Leviathan and the Air-Pump* understands Hobbes as defending that status of natural philosophy as "a causal enterprise [...] [that] as such, secured total and irrevocable assent" (Shapin and Schaffer 1985, 19). Indeed, Boyle may have seen himself as defending "the proper procedures of experimental philosophy against the beast of deductivism" (Shapin and Schaffer 1985, 176). The aim of the preceding two subsections has been to assess Hobbes's criticisms, regardless of Boyle's viewpoint of them, in light of Hobbes's broader works and view of method in natural philosophy.

166   HOBBES'S TWO SCIENCES

In Shapin and Schaffer's discussion of these criticisms (1985, 125ff), the following claim of Hobbes's figures centrally: "[…] ingenuity is one thing and method [*ars*] is another. Here method is needed" (Hobbes 1985, 347). Shapin and Schaffer understand this criticism as Hobbes making clear that he saw experimentalists like Boyle as denigrating the role of the philosopher. The philosopher inhabited a noble role, not something on par with that of an "apothecary" or a "gardener" (Hobbes 1985, 130). Shapin and Schaffer rely upon a possible etymological pun by Hobbes in his usage of 'ingenuity', perhaps meant to intimate both the sense of cleverness and related to a mill, as seen in the Old French *engin* derived from it (1985, 129–130).

However, Shapin and Schaffer's discussion neglect the crucial sentence that immediately follows the juxtaposition of ingenuity and method, a sentence that seems intended as a brief description of what counts as 'method': "The causes of those things done by motion are to be investigated through a knowledge of motion, the knowledge of which, the noblest part of geometry, is hitherto untouched" (Hobbes 1985, 347). We need not import notions of one version of the role of the philosopher as noble compared with another to understand Hobbes's criticism of Boyle and colleagues having mere ingenuity. Instead, Hobbes's claim should be understood in light of his view of the priority of geometry—geometry is the science that provides causal principles, which are the 'why' borrowed within a mixed-mathematical natural philosophy.

The issue of 'method' at stake in the *Dialogus Physicus* relates not to whether experiments should *ever* legitimately count within natural philosophical explanation. In this respect, the present account partly agrees with Shapin and Schaffer's view (Shapin and Schaffer 1985, 129). Instead, Hobbes should be understood as emphasizing the necessity of knowing the 'why' in advance of any experiment or experience, as well as the need for the principles of first philosophy that serve to constrain what can count as a possible cause. This is why in *De homine* X.5 (quoted in subsection 5.1.2) Hobbes claims that "nothing can be demonstrated by physics without something also being demonstrated *a priori*" (OL II.93). The 'why' in geometry must be known in advance of any experiment, and then principles are borrowed from geometry in an explanation.

Nevertheless, Hobbes's focus upon causal principles borrowed from geometry does not make the 'that' that one gains from experience or an experiment unimportant for doing natural philosophy. Instead, Hobbes claims that knowledge 'that' from both natural and civil history is "[…] very useful (no, indeed necessary) for philosophy [...]" (Hobbes 1985, 189; OL I.9; EW I.10).[31] Similarly, in *De homine* XI.10 Hobbes asserts that "[…] histories are particularly useful, for they supply the

---

[31] The role played by civil history within explanations of historical events is the subject of chapter 6.

experiences/experiments [*experimenta*] on which the sciences of the causes rest" (OL II.100).

Given Hobbes's commitments within the broader corpus surveyed in this chapter, I suggest that we understand his central complaint in the *Dialogus Physicus* as relating neither to the role of experiment or experience in natural philosophy nor to the ignoble nature of experimenting *qua* philosopher. Rather, the primary issue at stake for Hobbes is that proper method in natural philosophy must be followed at all times—the proper method begins with *a priori* work in first philosophy and geometry, deduces consequences in geometry-related motion and magnitude, and then borrows those causal principles to provide the reason 'why' in a natural-philosophical explanation.

Deduction will still play a role in a natural-philosophical explanation, but we should not, *pace* Shapin and Schaffer, view Hobbes or his system as the "beast of deductivism" against which Boyle mounts a defense. Hobbes does, of course, view the principles of geometry as following from first principles of body and motion, what Hobbes calls the "simplest conceptions" (OL I.62–63; EW I.70). However, a virtue of understanding Hobbesian natural philosophy generally, as well as the explanation of the nature of air in the *Dialogus Physicus*, as a mixed-mathematical science is that it removes the need to see deductive connections *between* the sciences of geometry and natural philosophy. The mixed-mathematics view of Hobbes's natural philosophy is thus able both to articulate the unity of geometry and natural philosophy as well as show that they are independent disciplines with differing epistemic standards. The natural philosopher will explain by means of syllogistic deductions, but when borrowing geometrical principles, the natural philosopher will treat the physical body as if it were a mathematical body. As a result, explanations in natural philosophy will attain suppositional certainty, not "total and irrevocable assent."

5.4 Conclusion

This chapter has argued that, rather than viewing natural philosophy as deduced from an armchair metaphysics, Hobbes held that "true physics" involved mixing the 'that' from experience with the 'why' borrowed from the causal science of geometry. The mixed-mathematics interpretation of Hobbes's natural philosophy makes sense of a tension in Hobbes's thought better than competing interpretations: how is it that in *De corpore* I.8 Hobbes can hold, on the one hand, that natural (and civil) history is not part of philosophy while stating, on the other hand, that it is "necessary" to doing philosophy (OL I.9; EW I.10; Hobbes 1981, 189)? The resolution of that tension becomes clear in Hobbes's actual practice of explanation in *De corpore* Part IV and in the *Dialogus Physicus*, as well in Hobbes's reflection on the role of experience in *Examinatio*. In these works, the present chapter has shown

168 HOBBES'S TWO SCIENCES

Hobbes borrowing and citing causal principles from geometry without a deduction playing the role of a bridge, while also using the two *a priori* principles from first philosophy to rule out explanations that were inconceivable.

Hobbes's assessment of experimental philosophy is evident through the agreement of both interlocutors in *Examinatio*: both agree that experimenters are "worthy of praise" because they "reveal new and marvelous works of nature" and instill a love for philosophy in others (OL IV.228). No deductivist of the sort that Shapin and Schaffer suggest would praise experimentalists in this way. Hobbes's proposal is straightforwardly that before engaging in experiments one must first determine what sort of causes are possible (i.e., by using the two *a priori* principles from first philosophy) and second examine the nature of motion and body (geometry, or the 'why', in which we consider "motion and magnitude in themselves and in the abstract" [OL I.314; EW I.386]). The former will rule out many causes as impossible, such as those that appeal to action at a distance or self-motion, while the latter will provide causes for which one has maker's knowledge (*scientia*). Only after this prior work has been conducted should one attend to one's own observations, or even experiments, to establish the 'that'. Thus, in any natural-philosophical explanation, Hobbes held that one should ideally have done the work of *De corpore* Parts II and III prior to doing the type of work he does in Part IV or the *Dialogus Physicus*.

Furthermore, the mixed-mathematics interpretation explains why Hobbes held natural philosophy in higher regard than mere prudence; although he did not accord it the epistemic certainty of *scientia*, the suppositional certainties natural philosophy provides place it higher than mere *cognitio*-based prudence. Chapter 6 argues that this same mixed-mathematical understanding of the relationship between experience and *scientia* grounded Hobbes's explanations of historical events.

6
Experience in Hobbesian Civil Philosophy and Civil History

Although Hobbes held that sense experience was the source of all conceptions, he placed little stock in conjectures that relied upon experience alone. Making conjectures based on experiences that have developed into prudence may often be useful in everyday life, but such inferences suffer from the limitations discussed in chapter 2: they are "very fallacious" and subject to error because of the "difficulty of observing all circumstances" (Hobbes 2012, 42; 1651, 10; cf. EL I.IV.7). The only hope for humans to escape their natural condition is to create something new, not merely based on sense-based *cognitio* but instead by creating something artificial for which they have scientific knowledge (*scientia*).

Such comments about the science of civil philosophy would seem to remove any need for the political scientist to know civil history. Indeed, in marking out the "subject" of philosophy as "every body of which any generation can be conceived and of which a comparison can be made after any consideration of it" (Hobbes 1981, 189; OL I.9; EW I.10), Hobbes excludes from philosophy a number of domains of inquiry, including theology, angelology, natural history, and political history. He places natural and political history outside of philosophy because "[. . .] such knowledge [*cognitio*] is either experience or authority, not reasoning," but in the same sentence he admits that they are "useful (no, indeed necessary [*necessarias*]) for philosophy" (Hobbes 1981, 189; OL I.9; EW I.10). Anyone familiar with Hobbes's oeuvre would deny that comments such as these are meant to dismiss history as inessential to his philosophical project. *Behemoth* and Hobbes's translations of Thucydides count among his "historical" works, but he also showed abiding interest in ecclesiastical history, as evident in *Leviathan* Part IV and *Historia Ecclesiastica* (1688).

If not part of philosophy but nevertheless useful for it—even necessary—what is the relationship between Hobbesian history and the science of civil philosophy? Although *Behemoth* and Hobbes's other historical works have received less attention than *Leviathan*, scholars have puzzled concerning the relationship between these two domains, with most seeing a tangential connection between them. For example, Patricia Springborg suggests that *Behemoth* is merely an "extended disquisition" of the law of the just and unjust (2003, 268), or differently

Hobbes's Two Sciences. Marcus P. Adams, Oxford University Press. © Marcus P. Adams 2025.
DOI: 10.1093/9780198924715.003.0006

170   HOBBES'S TWO SCIENCES

put, it is a contextualization of the "science of *just* and *unjust*" (2003, 287). Similarly, Royce MacGillivray (1970, 179) proposes that *Behemoth* gave Hobbes a chance to "apply his philosophy of the state" to a catastrophic situation, and William Lund (2003, 59–60) suggests that Hobbes might have seen *Behemoth* as a "realistic analogue of the hypothetical state of nature" that could provide a "vivid and rhetorically powerful illustration."[1] Vaughn (2002, 95–96) criticizes views that understand *Behemoth* as providing an illustration of Hobbes's civil philosophy or as aimed at explaining the civil war itself, instead arguing that *Behemoth* generally aims to show the nature of Hobbesian political pedagogy. My aim is not to offer an account of what the *Behemoth* is generally but instead to examine why Hobbes thought it was legitimate to borrow causal principles from civil philosophy within certain parts of historical *explanations.*

The answer one gives to the question of how history relates to philosophy depends upon how one understands the nature of Hobbesian history and historical writing. Perhaps 'history' is a mere collection of experiences that are remembered either by individuals or by groups through oral or written transmission. Let us call the individual who practices this the mere historian who produces natural history or political history. Were this Hobbes's entire view of the nature of history, he ought to have a dim view of its epistemic potential since, even if an account is well supported in its findings, say, by drawing upon a large body of eyewitnesses or historical records, it will nevertheless suffer from the limitations of mere prudence (*cognitio*).

An alternative understanding of Hobbes's historical activity emphasizes its aim to *explain* the historical past: it seeks to explain *why* a particular historical event happened. For such an endeavor, clearly the facts are essential, but something more is required beyond those facts to issue a judgment for why an event occurred—to understand its causes. Let us call an individual engaged in this activity a historiographer. In this chapter, I argue that the Hobbesian historiographer engages in the same type of activity described in chapter 5 for the natural philosopher. Just like the

---

[1] Kraynak (1982) offers a more sophisticated account of this relationship by suggesting that *Behemoth* shows the "problem" to be solved while the science of civil philosophy provides the "solution." Kraynak proposes that *Behemoth* is the "true beginning" of Hobbes's civil philosophy insofar as it shows that the deficiencies of prior civil societies are due to an improper understanding of the nature of opinion and justice. As such, according to Kraynak's view, this diagnosis provides inspiration for civil philosophy: "[...] since history shows that the problem is opinion and reasoning from opinion, Hobbes writes treatises that methodologically analyze the elements of 'opinion'" (1982, 841; cf. Kraynak 1990, 69ff). Kraynak's view is not without problems. For example, if history played the role for Hobbes that Kraynak suggests, then it would serve as a sort of pre-theoretical inquiry that could inform the starting points for science; however, this would be exactly the sort of methodological primacy given to experience and prudence that Hobbes assails in the work of Boyle and others in the Royal Society. Furthermore, although Kraynak's account captures some of the sense in which history could be "useful (no, indeed necessary) for philosophy" (Hobbes 1981, 189; OL I.9; EW I.10–11), it does not show *why* Hobbes thought he could legitimately use principles from civil philosophy within history.

EXPERIENCE IN HOBBESIAN CIVIL PHILOSOPHY   171

natural philosopher ideally mixes the 'that' and the 'why' in explanations of natural phenomena, I argue that the Hobbesian historiographer first establishes that some sequence of events happened (the 'that') and then borrows a possible cause (the 'why') from the science of civil philosophy. Since possible causes borrowed from civil philosophy are cases of *scientia*, using them results in an explanation more epistemically certain than mere history (i.e., prudence).

Rather than understanding Hobbes's statement that, even though not on par with philosophy, political history is "useful (no, indeed necessary) for philosophy" (Hobbes 1981, 189; OL I.9; EW I.10–11) as asserting a need for *preparation* in history by a philosopher, the view I defend shows that in any explanation of the historical past, both political history and civil philosophy play essential roles to make historiography. Seeing this connection between civil philosophy and historiography illuminates the source of tension in Hobbes's seemingly paradoxical statements about the place of history. In short, it shows that the explanations in chapter 45 of *Leviathan* Part IV are structurally analogous to those in the natural philosophy of *De corpore* Part IV, and so likewise are the explanations in the historiographical dialogue *Behemoth* and the natural-philosophical *Dialogus Physicus*.[2] When Hobbes provides such causal explanations of historical events—what I am calling historiography—that activity counts as philosophizing just like natural philosophy does.

The chapter proceeds in three stages. First, I contrast Hobbes's early comments on experience and historiography in the translation of Thucydides's *History of the Peloponnesian War* (1628) and *Elements of Law* (1640) to the later distinctions Hobbes makes in *Leviathan*, not present in *Elements of Law* or *De cive*, regarding experience in confirmation and disconfirmation. Second, I show how the model of "mixing" the 'that' with the 'why' explains Hobbes's historiographical work in *Behemoth* Dialogue 1 and in *Leviathan* 45 in Part IV "The Kingdom of Darkness." Third, I address two potential worries with the structural analogy that I make between natural philosophy and historiography— that explanations in Hobbesian historiography seem exclusively negative and that historical events are unrepeatable.

---

[2] Springborg (2003, 268) connects the dialogue format of *Behemoth* to *Dialogue between a Philosopher and a Student of the Common Laws of England* (1681), the *Historia Ecclesiastica*, and the 1668 Appendix to the Latin *Leviathan*. In these dialogues, Springborg suggests, Hobbes strengthens the arguments of *Leviathan* by using legal case methodology, ecclesiology, and historiography, respectively. However, in neglecting the *Dialogus Physicus* (discussed in chapter 5), Springborg misses a point of connection: Hobbes is using the popular dialogue format to offer a simulated mixed-mathematics explanation. My aim is not to argue that *Behemoth* consists entirely of historiography any more than my aim was to argue that the *Dialogus Physicus* was entirely mixed-mathematical natural philosophy. Each of these dialogues can have different aims at different points; my goal is only to show that viewing *Behemoth* alongside *Dialogus Physicus* illuminates Hobbes's explanatory strategy in borrowing causal principles to explain some historical or experimental phenomenon.

172 HOBBES'S TWO SCIENCES

6.1 History and Experience: Confirmation and Disconfirmation

6.1.1 Hobbes's Seemingly Conflicting Views on History

From his earliest writings through the completion of the three volumes of the *Elements of Philosophy*, Hobbes recognized the limitations of relying exclusively on experience. For example, although in *Elements of Law* Hobbes asserts that "[o]riginally all conceptions proceed from the actions of the thing itself, whereof it is the conception" (EL I.2.2), he warns that "[e]xperience concludeth nothing universally" (EL I.4.10). Admitting that many identify a greater number of experiences—someone who one who takes "signs from experience"—as what sets apart the wise person from the unwise, Hobbes denies that this assessment is legitimate because the signs from experience are merely conjectural (EL I.4.10).

In this early work, Hobbes had not yet developed his account of *scientia* as maker's knowledge (discussed in chapter 3), as is evident from his description of the "two kinds" of knowledge: the first kind is sense and remembrance of sense; and the second is "science, or knowledge of the truth of propositions, and how things are called, and is derived from understanding" (EL I.6.1). Unlike the contrast between "sense and Memory" and "science" in *Leviathan* 6 (2012, 72; 1651), where the latter includes causal knowledge that imparts the ability to produce like effects from like causes, in *Elements of Law* Hobbes is explicit that both kinds of knowledge are "but experience" (EL I.6.1). This apparent reduction of all knowledge, even "science", to experience is clear from two claims Hobbes makes: first, that 'history' is simply the register of remembrance; and second, that the registers of the sciences are simply records of how to use names properly in language.[3] Reading a historical text, then, would be no different from discussing the explanation of some natural phenomenon with an experienced individual; the best for which one could hope from such an encounter would be an increase in one's own experiences that could lead to prudence.

---

[3] This connection of science to knowing the proper use of names in language has led some scholars to see Hobbes as a type of conventionalist, where doing science is more about syllogistic reasoning from stipulative, arbitrarily chosen definitions (e.g., Martinich 2005, 165; McNeilly 1968, 67; Shapin and Schaffer 1985, 149–150). Although Hobbes had not developed the maker's knowledge framework I discussed in chapter 3, I have argued elsewhere that there are reasons to resist understanding him as a conventionalist even in this early period. For example, although Hobbes emphasizes the role of language in *Elements of Law*, and identifies truth as a property of propositions, he articulates "evidence" as "concomitance of a man's conception with the words that signify such conception in the act of ratiocination" (EL I.6.3). Evidence is a necessary condition for truth—like "sap is to the tree"—so "science" must have a connection to conceptions received in sensation. As Hobbes notes, even parrots can "speak" truths, but they cannot know it because they lack evidence. For more detailed arguments against the conventionalist interpretation of Hobbes and on his understanding of 'evidence', see Adams (2014a; 2014b, 55–60).

EXPERIENCE IN HOBBESIAN CIVIL PHILOSOPHY   173

Later in *Elements of Law*, Hobbes worries about the limits of history as the register of human remembrances. He worries that although words function as "signs of another's opinions and intentions," the meaning of ambiguous words will depend on one's context and audience. Living communicators sort this out by asking questions or by considering a speaker's time and place, but a reader's ability to do this is limited when reading history. Hobbes recognizes that

> it must be extreme hard to find out the opinions and meanings of those men that are gone from us long ago, and have left us no other signification thereof but their books; which cannot possibly be understood without history enough to discover those aforementioned circumstances, and also without great prudence to observe them. (EL I.13.8)

This worry about understanding a writer's meaning, combined with Hobbes's reduction of history to the mere register of sense, makes clear that history is at best epistemically limited. Nevertheless, Hobbes does later appeal to history in *Elements of Law*, for example, when asserting that "all the ancients have preferred monarchy before other governments" both in their opinions and in their customs (EL II.5.3).

The limited utility accorded to history in *Elements of Law*, both in terms of its output of mere prudence and the difficulty of knowing what past authors meant by the words they used, stands in contrast to Hobbes's comments a little over a decade earlier in his translation of Thucydides's *History of the Peloponnesian War* (1629). There Hobbes extols the use of history by proclaiming its pedagogical benefits, which would have been recognizable to anyone familiar with a humanist understanding of history (Raylor 2018, 71):

> For the principle and proper work of history being to instruct and enable men, by the knowledge of actions past, to bear themselves prudently in the present and providently towards the future: there is not extant any other (merely human) that doth more natural and fully perform it, than this of my author. (EW VIII.vii)

The instruction for which Hobbes praises Thucydides relates to method: rather than inserting dialogues aimed at teaching the reader, or explicitly conjecturing the causes for an event, Hobbes praises Thucydides's ability to make the "auditor a spectator" (EW VIII.viii).[4] Thucydides's "choice of matter" and ordering with "judgment" accomplishes this feat and this, Hobbes suggests, is the reason why he is taken to be the "most politic historiographer that ever writ" (EW VIII.viii).[5]

---

[4] Evrigenis (2006, 307–308) connects this praise to Hobbes's account of the origin of ideas in sensation developed more fully in later works. Sowerby (1998, 153) makes a similar connection.

[5] Hobbes later reaffirms the role of judgment in "a good history" in *Leviathan* 8 (2012, 106; 1651, 33).

174   HOBBES'S TWO SCIENCES

In defending Thucydides against Dionysius of Halicarnassius's criticisms, Hobbes emphasizes that the historiographer should both discover and describe the stated causes of war (the "public and avowed cause of this war") as well as conjecture about the "true and inward motive" behind actions (EW VIII.xxvii). Both aspects, Hobbes argues, are "within the task of the historiography, no less than the war itself" (EW VIII.xxvii). Historiography should display the pretext for the war as well as the conjectured "inner" causes, concerning which method Hobbes suggests "a more clear and natural order [than Thucydides's] cannot possibly be devised" (EW VIII.xxvii). Hobbes is clear that this work of the historiographer provides the reader with a type of experience akin to having witnessed the events themselves and being able to draw one's own conclusions about the causes responsible for them (EW VIII.viii).

How can we reconcile these two aspects of Hobbes's thought regarding the status of historiography? Whereas in 1629 we find Hobbes extolling the benefits of historiography, in 1640 he has nearly dismissed it as nothing more than recorded experiences and expresses worries about the difficulties associated with understanding the words used by those from different times and places. Given the gap in time between his translation of Thucydides and the *Elements of Law*, an unsurprising strategy that Hobbes scholars have employed suggests that Hobbes simply changed his mind; call this the developmental strategy for understanding Hobbes (discussed earlier in chapter 1). Hobbes the "humanist" was responsible for translating Thucydides, but eventually Hobbes the "philosopher" realized that history could never achieve the universality that is possible when one starts from general principles. As Raylor puts it, by the time of *Elements of Law*, Hobbes held that "experience and prudence, and therefore history, cannot be considered productive of true knowledge" (2018, 203).

Schuhmann likewise employs this strategy and suggests that Hobbes changed his mind because history could never provide "an infallible rule of action" (2000, 14); only philosophy could do this.[6] As support, Schuhmann draws upon Hobbes's criticisms against using ancient history for learning about civil policy: "[. . .] one of the most frequent causes of [rebellion against monarchy], is the Reading of the books of Policy, and Histories of the antient Greeks, and Romans" (2012, 506; 1651, 170). Schuhmann is right to draw attention to Hobbes's caution regarding the use of ancient history, but Schuhmann reaches this conclusion too quickly. Hobbes's

---

[6] On shifts in Hobbes's thought about the sources of knowledge, see Raylor (2018, 194–195). On the apparent rejection of history in Hobbes's later works, see Raylor (2018, 202–203) and Skinner (1996, 259ff). Steinmetz (2021) draws attention to Hobbes's continued interest and work in translation in *both* the humanist and philosophical phases that some scholars have identified. Although Steinmetz's claim that Hobbes found in Thucydides the "universal principle" that violence results when no common authority is present (see Steinmetz 2021, 91) cannot be substantiated by Hobbes's actual epistemology of *scientia*, Hobbes's continued interest in translation and history throughout his career presents a challenge for an understanding of Hobbes's thought that sees shifts in his thought as discontinuous with one another.

complaint is not with the reading of ancient history as such but rather with those who fail to see the many problems resulting from those ancient forms of government, as Hobbes continues: "Not considering the frequent Seditions, and Civill warres, produced by the imperfection of the Policy" (2012, 506–508; 1651, 170).

In trying to explain how Hobbes could praise history in 1629 while appearing to reduce it to the mere record of experience by the 1640s, the developmental strategy runs afoul of the fact that Hobbes's interest in history abided throughout his later writing period, especially with *Behemoth* and *Historia Ecclesiastica*. Furthermore, Hobbes himself declares in 1655 in *De corpore*, as mentioned already, that both civil and natural history are "useful (no, indeed necessary [*necessarias*]) for philosophy" (1981, 189; OL I.9; EW I.10–11). The developmental strategy has difficulty explaining the *necessity* of civil history and natural history if Hobbes the mature "philosopher" had left them behind. Rather than imposing categories of Hobbes's development on the way to his mature thought in *Leviathan* and the *Elements* trilogy, I suggest that throughout Hobbes's corpus we can trace a role for civil history that is analogous to the role played by natural history: in an explanation of some event, civil history provides the historiographer with the 'that', while civil philosophy provides the reason 'why'.[7] Hobbesian historiography differs from the Thucydidean ideal that Hobbes earlier espoused, where the reader "may from the narrations draw out lessons to himself" (EW VIII.viii) or where the narrative "may secretly instruct the reader" (EW VIII, xxii), but this abandonment is unsurprising given Hobbes's many criticisms of the limits of prudence and his view that the everyday person is unable to know the causes of events, whether natural or civil, simply from experience.

Brief examples of this use of civil history to demonstrate *that* some event *B* happened following some preceding event *A* abound throughout Hobbes's works from *Elements of Law* onward, and Hobbes often uses either a Law of Nature or some other principle from civil philosophy to provide the reason *why*. For example, when discussing the Law of Nature "*That men allow commerce and traffic indifferently to one another*" (EL I.16.12), Hobbes mentions the violation of this Law as preceding the Peloponnesian War. The support Hobbes provides for the law, however, is not founded in the historical case; as support, he argues that whenever someone allows commerce and trade to one person but not to another, they declare hatred for the latter person, and "to declare hatred is an act of war." Hobbes uses a historical fact to show that a violation of this law has occurred in the past (the Athenians did not allow the Megareans to use their ports and markets) and shows that the war followed that disallowance (the 'that'), but the historical record itself

---

[7] My account thus opposes Schuhmann's attempt to distinguish civil and natural history by claiming that "[. . .] where natural philosophy is to build on the reliable foundation of natural history, political philosophy need not have recourse to distant ages in order to find its foundation" (2000, 15). As is clear in chapter 5 and in the present chapter, Hobbes does not hold that history, whether natural or political, provides a "foundation" for either political philosophy or natural philosophy.

176   HOBBES'S TWO SCIENCES

does not provide the reason *why*. He invokes causal language when he asserts that "[...] upon this title was grounded[8] the great war" (EL I.16.12); in other words, the Law of Nature explains the sequence of events from A to B (the 'why').

Another such brief example is when Hobbes provides the principle that a monarchy is less likely to dissolve because a monarchy is a unity. The argument he makes is conceptual:

> For where the union, or band of a commonwealth, is one man, there is no distraction; whereas in assemblies, those that are of different opinions, and give different counsel, are apt to fall out amongst themselves, and to cross the designs of commonwealth for one another's sake: and when they cannot have the honour of making good their own devices, they yet seek the honour to make the counsels of their adversaries to prove vain. And in this contention, when the opposite factions happen to be anything equal in strength, they presently fall to war. (EL II.5.8)

Hobbes's argument is that one individual simply has no distractions from the competing opinions or priorities present in multi-person governing bodies. However, when factions with different opinions are nearly equal in strength (and thus possess hope of gaining their ends), a virtual state of nature recurs, and war ensues. As historical examples of such conflict, Hobbes cites Athens and Rome because they allowed "affairs of state [to be] debated in great and numerous assemblies." This brief historical example could not serve as evidence for Hobbes's claim about monarchy since a chapter earlier he clearly states "[e]xperience concludeth nothing universally" (EL I.4.10). Furthermore, Hobbes's reasoning is not based upon facts about human psychology and group behavior. His view allows that groups are capable of unity when their members do not have different opinions (or when they act simply to choose those who will handle the affairs of state). Furthermore, even if members of a group do not have different opinions, it is not necessary that this results in disagreement; Hobbes says only that such groups are "apt" to do so. In contrast, for a single individual his claim about unity follows necessarily: "there is no distraction."

It may seem tempting to understand such brief examples as mere *illustrations* of Hobbes's conceptual points, perhaps merely for rhetorical sake. However, my suggestion is that even brief examples like this show *that* some sequence of events has occurred, and then the reason *why* is drawn from the science of civil philosophy. The reason *why* Athens and Rome dissolved and faced many internal conflicts is *because* they had divided government. The cause is not drawn from a historical induction by observing similar associations in Humean fashion, but rather the

---

[8] Hobbes elsewhere uses 'ground' as a synonym for 'cause', e.g., when explaining why in dreaming humans do not "wonder [...] in their dreams at places and persons, as they would do waking" (EL I.3.9).

conceptual claim about unified government follows from concepts like 'unity' and 'one man'. As Hobbes puts it, even after a dissolution occurs and war breaks out, "necessity teacheth both sides" that they need an "absolute monarch" both to defend them from the other faction and to keep their own faction at peace (EL II.5.8).

So far, my aim has been to show that far from abandoning history for philosophy, Hobbes developed a more nuanced account of the relationship between the two domains. While in *Elements of Law* Hobbes is clearly developing his understanding of the Laws of Nature in earnest, historical facts play an essential role in that work because they provide opportunities for Hobbes to provide explanations of events by using the laws.[9]

6.1.2 History and *Scientia*: Confirming and Disconfirming in Later Works

By the time Hobbes published *Leviathan* in 1651, he had been working for some time on the material that would later be included in *De corpore*. Although Hobbes did not publish the latter until 1655, he had the building blocks for his maker's knowledge epistemology already in place in the late 1630s and early 1640s, and contemporaries such as Robert Payne and Charles Cavendish were aware of the epistemic foundation of "simplest conceptions" that Hobbes later went on to defend in print.[10] In these later developments, Hobbes makes clear that *scientia* provides bearers with causal knowledge that they could use to make something, such as a geometrical figure like TRIANGLE or PEACE. Before I provide two examples of Hobbesian historiography in the next section (section 6.2), the present subsection briefly describes two additional relationships between history, considered generally, and *scientia* in *Leviathan*: confirming or disconfirming the results of *scientia* by using history. These concern the role of history within *scientia*: I argue that Hobbes approves of the use of historical facts for confirming the results of *scientia* but disallows the use of history in attempts to disconfirm those results.

After demonstrating in the first part of *Leviathan* 13 that humans in their natural state would be driven by their natural equality and their passions to a state of

---

[9] This approach is at odds with claims that Geoffrey Vaughn has made about the lack of engagement in *Behemoth* with the Laws of Nature. Vaughn asserts that the "state of nature, the laws of nature, and the move to civil society do not play a central, if any, role in *Behemoth*" (2009, 178). Rather than seeing the "conclusions" of *Behemoth* explained by appeal to Hobbes's civil philosophy, Vaughn asserts that they are explained "in terms of history" (2009, 179). Vaughn's approach assumes that the aim of civil philosophy can only be to argue *for* conclusions about sovereignty, but according to the account of the Laws of Nature I have offered in chapter 4, the Laws could also be used to diagnose failures to make PEACE (in just the same way that neglecting to follow instructions for how to make SQUARE could be used to diagnose one's failure in construction). Furthermore, as I discuss below, there are clear uses of the Laws of Nature within the text of *Behemoth*.

[10] Elsewhere, I have defended this claim by situating Hobbes's epistemology of simplest conceptions within the context of his Objections to Descartes's *Meditations* (see Adams 2014a).

178 HOBBES'S TWO SCIENCES

war, Hobbes considers two potential objections to the "inference" that he has just made: that the portrayal of humans in this state is unrealistic and that no such state has ever existed. Rather than encouraging the reader to re-read more carefully the demonstration just provided, Hobbes does something that, given his earlier demarcation of "science" from mere sense and memory in *Leviathan* 5 (2012, 72; 1651, 21), may seem strange. He acknowledges that the reader may wish to "have the same confirmed by Experience" (2012, 194; 1651, 62). The 1668 Latin edition is more emphatic, stating that the conclusion reached "has been clearly inferred from the nature of the passions, and is also agreeable to experience" (2012, 195).

The two appeals to experience that Hobbes makes are well known. First, Hobbes suggests that every person's behaviors in locking up doors and chests "accuse mankind" in the same way that the demonstration has just done (2012, 194; 1651, 62). Second, although Hobbes admits that there was never any general natural state of humans, there are places he suggests "where they live so now" (2012, 194; 1651, 63). To this second appeal, he adds that sovereigns themselves adopt a "posture of War" against one another, which is clear from their continual need to arm themselves and construct ever better defenses (2012, 196; 1651, 63). These two examples have received much attention in themselves, but my interest is in the status of these appeals within the science of civil philosophy.[11] Hobbes's description of the person needing to use experience to aid them—someone who "has not well weighed these things" and who thus cannot "trust" the inferences—makes clear that this is not his preferred route. Nevertheless, his suggestion that experience can *confirm* the output of *scientia* suggests that this is more than mere rhetoric.

Hobbes's approval of the use of history for personal confirmation contrasts starkly with his complete rejection of its use for any attempt to *disconfirm* the output of a scientific demonstration. Although *Leviathan* 5 portrays prudent individuals with "no *Science* [1668: *Scientia*]" as better than those who reason badly or trust those who do (2012, 74; 1651, 21), Hobbes clearly thinks that when experience and scientific knowledge disagree, the latter must be preferred. Given his distinction between "reason" as a cognitive activity in which humans engage and "right reason" as an objective standard, Hobbes allows for human error in scientific demonstrations. For example, although humans err in arithmetical calculations, Hobbes emphasizes there is an objective standard: "Arithmetique is a certain and infallible Art" (2012, 65; 1651, 18). Likewise, although individual humans err in reasoning, Hobbes emphasizes that as an objective series of steps in a demonstration "Reason it selfe is always Right Reason" (2012, 65; 1651, 18).[12]

---

[11] I discuss confirmation from experience in the context of Hobbes's view of demonstration and analysis and synthesis in chapter 4.

[12] Although Hobbes's distinction between the objective correctness of the inferences within a science and the errors of finite humans when attempting to make those inferences is intuitive, he recognizes that, practically speaking, humans will frequently need to consent to a third party to act as judge in cases of dispute.

This distinction makes clear that the inferences within a science themselves ground the certainty of the conclusions, not the speaker or group who asserts them. As Hobbes puts it, neither a single person's reason nor the "Reason of any one number of men" is what lends certainty; unanimous approval has no bearing on objective certainty (2012, 65; 1651, 18).

This distinction between individual or group error in reasoning and the objective nature of scientific demonstration is evident in Hobbes's later discussion of what he calls the "greatest objection" to his account of sovereignty. Although he elsewhere praises the benefits of prudence, as discussed in chapter 2, Hobbes rejects any attempt to criticize his account of sovereignty on that basis. Even universal consent provides no ground for rejecting the results of a scientific demonstration. He argues that even if "in all places of the world, men should lay the foundation of their houses on the sand, it could not be thence inferred, that so it ought to be" (2012, 320–322; 1651, 107).[13] Unlike prudence-based practice, which he compares to "Tennis-play," Hobbes argues that the "skill [1668: *scientia*] of making, and maintaining Common-wealths" consists in "certain Rules," just like those followed in arithmetic and geometry. In sum, any attempt to use history, or experience generally, to *disconfirm* the results of an inference within *scientia* is illegitimate. The next two sections provide examples of Hobbesian historiography and show Hobbes using historical facts to establish *that* some sequence of events happened and then appealing to *scientia* to explain *why* they turned out as they did. Section 6.2 examines Hobbes's explanation of the Civil War, and 6.3 focuses upon his explanation of the bondage of Christian subjects in *Leviathan* Part IV.

6.2 The "Seeds" of the Civil War in *Behemoth*, First Dialogue

Hobbes wrote the *Behemoth*, like *Dialogus Physicus*, in dialogue form. Hobbes seems to have chosen this genre when addressing controversial topics and when wanting to defend his own views indirectly. Hobbes's contemporary John Wallis humorously suggests that Hobbes wanted to give the appearance of each interlocutor agreeing with or coming to agree with Hobbes's own view: "[. . .] [Hobbes] found a middle course, by way of Dialogue, between A and B, (*Thomas* and *Hobs*;) Wherein *Thomas* commends *Hobs*, and *Hobs* commends *Thomas*, and both commend *Thomas Hobs* as a third Person; without being guilty of self-commendation" (Wallis 1662, 15; quoted in Hobbes 2010, 19).

While scholars have typically treated Hobbes's dialogues with a more nuanced approach than Wallis did, Hobbes's use of his own material from other works within these dialogues has sometimes still been seen as a form of thinly veiled

---

[13] I discuss Hobbes's response to the objection from experience when distinguishing between prudence and *scientia* in chapter 3.

180   HOBBES'S TWO SCIENCES

self-service. For example, in his introduction to the Clarendon edition of *Behemoth*, Paul Seaward assesses interlocutor A's "solution" to gesture toward the "Rules of *Just* and *Vnjust*" as nothing more than "Hobbes's plea for the adoption of his own philosophy, and perhaps himself, as the key to securing permanent civil peace" (Seaward in Hobbes 2010, 24). While Hobbes's defense of his system is in the background, as it is in his criticisms of Boyle's experimental philosophy in the *Dialogus Physicus*, the present section argues that Hobbes's explicit appeals to the science of civil philosophy within *Behemoth* function as the borrowing of causal principles. The use of causal principles as the 'why' in combination with a narration of the facts (the 'that') makes it possible for the explanation of the Civil War to be more than simply a recounting of events, which as mentioned already would be merely *cognitio* that is "either experience or authority, not reasoning" and thus excluded from philosophy (Hobbes 1981, 189; OL I.9; EW I.10–11). Although more certain than experience alone, this mixed explanation—what I am calling Hobbesian historiography—is less certain than the conclusions of civil philosophy.

Behemoth is divided into four dialogues. In the dedication to the work, Hobbes identifies the First Dialogue, which will be the focus of the present section, as providing "the seed of [the Civil War], certain opinions in Diunity and Politicks" (Hobbes 2010, 106). Hobbes elsewhere uses 'seed' to refer to a cause, for example, when describing the "naturall cause of religion" in *Leviathan* 12 (discussed in subsection 6.3.1). The First Dialogue begins by prioritizing the role of sense. Interlocutor A suggests that if a person had been atop the "diuells mountain," a likely reference to the great height to which Jesus was taken by Satan to observe various kingdoms of the world, then that person "should have looked vpon the world, and obserued the actions of men, especially in England, and might have had a prospect of all kinds of Iniustice, and all kinds of Folly that the world could afford [...]" (2010, 107). Speaker B expresses a desire to be placed upon "the same mountain." This starting point seems to be an allusion to Hobbes's earlier praise concerning Thucydides's method in which the "auditor [became] a spectator" (EW VIII.viii).

The opening proposal that interlocutor A offers, as Hobbes's mouthpiece, for the cause of the War is that the subjects ruled by Charles I were corrupted by seducers in the following seven groups: (1) Presbyterian ministers; (2) Papists; (3) so-called independents, such as Anabaptists and Quakers; (4) individuals who were enthralled with Ancient Greek and Roman forms of government; (5) city-dwellers who saw the financial prosperity brought on by rebellions elsewhere; (6) people who desired war after being in financial disarray; and (7) individuals who were ignorant of their duty and of the need for a sovereign (Hobbes 2010, 109–110). That Hobbes would offer many factors like this is consistent with his understanding of "entire cause" (*causa integra*), discussed in chapters 3 and 4, where a cause is "the aggregate of all the accidents both of the agents how many soever they be, and of the patient, put together" (OL I.107–108; EW I.121–122). However, at this level

EXPERIENCE IN HOBBESIAN CIVIL PHILOSOPHY   181

of generality, A has not yet provided an informative or surprising explanation. B responds that the descriptions provided so far make it difficult to "imagine how the King should come by any meanes to resist them" (2010, 111).

A replies that B will be "better informed" by the narration that follows. A's use of 'narration' to describe what follows in the text brings a further connection to the praise he offered of Thucydides's narrative method.[14] Hobbes uses 'narration' and other forms of that word elsewhere, such as in the title of *An Historical Narration concerning Heresy and the Punishment Thereof* (1680). The *Historical Narration* sought to redefine 'heresy' by linking it to its etymology ("different opinions") and tracing through different contexts how the word had been used. Elsewhere, in *Leviathan* 33, Hobbes denies that Jonah was the likely author of the biblical book named for him because it was "not properly a Register of his Prophecy [...] but a History or Narration of his frowardnesse and disputing Gods commandements" (2012, 598; 1651, 202). The Latin removes the disjunction by identifying the book of Jonah simply as a "historical narrative [*Narratio historica*]" (2012, 599; 1668, 179). I take speaker A's reference to parts of what follows in *Behemoth* as a "narration" to refer to the presentation of the facts (i.e., the 'that'), which he interrupts at several points by borrowing a causal principle from the Laws of Nature in civil philosophy. I discuss two such instances: the Seventeenth Law of Nature and the Third Law of Nature and the "sum" of the Laws. In both instances, the *explanandum* is the Civil War. The explanation for the War that speaker A offers has two parts: the 'that' is shown by reference to historical events, and the *explanans* (the 'why') is drawn from civil philosophy.

6.2.1 The Seventeenth Law of Nature in *Behemoth*

After A's promise to provide a narration, B requests to "know first the seuerall grounds of the pretences" of the Pope and the Presbyterians that caused them to think that they had any right to govern subjects (2010, 111). Here B is requesting to know *why* those in groups 1 and 2 above took themselves to have any legitimate authority. A proceeds by outlining why Papists took the Presbyterian claim to authority as illegitimate, basing their claim upon texts in the Vulgate used as evidence that priests served as judges in Israel, as well as texts describing Christ authorizing the Apostles, chiefly Peter, to be teachers who must be obeyed. A further describes the Roman Church's assertion that Peter's authority transferred through the line of Papal succession (2010, 111–112). A then repeats some of the argument from *Leviathan* 42 against Bellarmine's claim that spiritual power implies indirect temporal power; that is, even if a Pope does not have direct authority to judge and

---

[14] On the Baconian background of history and narrations, see Raylor (2018, 68–69).

182   HOBBES'S TWO SCIENCES

punish those who break "Ciuill Laws," he does have authority over this arena in-sofar as it can lead to the "hindrance or aduancement of Religion and good man-ners" (2010, 113).

Although this distinction between temporal and spiritual power seemed to re-serve some authority for civil sovereigns, A asserts that in the end it left "None or very little" (2010, 113). Beyond excommunicating individuals or nations, Popes appealed to the Roman Church's spiritual powers to require sovereigns to exempt church priests and other offices and church activities from criminal prosecution or taxation. Likewise, A reports that Popes claimed the power of "absoluing subiects of their duties, and of their oaths of fidelity to their lawfull Soueraignes" any time the Pope accused sovereigns of heresy (2010, 115). This was accomplished by claiming that canon law had a higher status than civil law. Interlocutor A continues by re-defining the meaning and use of 'heresy' and by tracing its use through var-ious church councils and by Popes (2010, 116–119).

Speaker A next describes the enforcement of doctrines like marriage regulation for priests, which placed civil sovereigns in a dilemma about whether to have an heir or be a priest, the necessity of auricular confession for salvation, and the doc-trine of transubstantiation (2010, 123–126). At this point in the dialogue, speaker B tries to interject with worries about identity and transubstantiation by asserting that "[i]t seems then that Christ had many Bodies, and was in as many places at once as there were Communicants" (2010, 126). A refuses to engage with this worry and emphasizes that the present is a "narration, not [...] a disputation" (2010, 126). The narration continues to trace the influence of the Roman Church, as well as of universities, in England, with B wondering how Henry VIII could "so vtterly ex-tinguish the Authority of the Pope in England, and that without any Rebellion at home, or inuasion from abroad" (2010, 130).

The answer A provides relies upon five details from the historical record, but my focus will be only on the second: "[...] the Doctrine of *Luther* beginning a little before, was now by a great many men of the greatest iudgments so well receiued, as that there was no hope to restore the Pope to his Power by Rebellion" (2010, 131). The precise way in which Luther's "Doctrine" was part of the cause of the Civil War becomes apparent later when A addresses the conflict in England be-tween Reformed believers generally and Papists. B wonders how it was pos-sible that the Presbyterians were able to gain power when they were simply "so many poor Schollers" lacking in financial resources, especially compared to the Roman Church (2010, 134). A locates their power not in something intrinsic to Presbyterianism itself but rather in the Protestant Reformation generally; with the Reformation, individuals became judges of right and wrong because they had ac-cess to the scriptures in their vernacular.

Although there are many anti-papist invectives throughout *Behemoth* and Hobbes's other works, A next positively compares the Pope's pre-Reformation role to Moses's role as the one who communicated directly with God: "For the Pope did concerning the Scriptures, the same Moses did concerning mount Sinai, Moses suffered no man to go vp it to hear God or speake or gaze vpon him, but such as he himselfe tooke with him" (2010, 135). Like Moses's exclusive access to God on Sinai, prior to the Reformation Popes held exclusive access to God not only by their office but, more importantly, because of the inability of everyday believers to read the Scriptures independently.

A agrees with B's assessment of Moses that such a limited access, accompanied by a reliance upon Moses as judge, was something done "very wisely and according to Gods owne commandement" (2010, 135). The events that followed the translation of the Scriptures into believers' native languages, A suggests, are an instance of what would have happened had God not commanded Moses to be the sole point of contact:

> For after the Bible was translated into English, euery man, nay euery boy and wench that could read English, thought they spoke with God Almighty and vnderstood what he said [...]. (2010, 135)

The result is that the "reuerence and obedience" that was due to the Reformed church was "cast off" because "euery man became a Judge or Religion, and an Interpreter of the Scriptures to himselfe" (2010, 135). The immediate worry B raises is that the Church of England itself had allowed translation of the Scriptures. If the Church of England had not intended the translated Scriptures to guide believers, then, B states, they would have "kept it, though open to themselues, to me seald vp in Hebrew, Greek, and Latine" (2010, 135).

A admits that the "lycence" given after the Reformation by translations that became available in vulgar languages was responsible for the birth of many sects, but A suggests that these divisions were "hidden till the beginning of the late Kings reign." However, once they were evident, A claims that they did "appear to the disturbance of the Commonwealth" (2010, 135). Following this linkage between individual interpretive license and the disturbance of the Commonwealth's peace, A suggests that they should "return to the story" at hand. A's diagnosis of the problems related to group 1 (Presbyterians) and group 2 (Papists) in the widespread availability of native-language translations would have been unsurprising to Hobbes's readers for whom the Reformers' exhortation to return *ad fontes* was known. I suggest that locating this as an issue related to *private judgment* makes clear that this is a causal claim related to the maintenance of peace from Hobbes's civil philosophy.

Private judgment figures centrally in Hobbes's well-known discussion of the right of nature in *Leviathan* 14. The right of nature is each person's liberty to "use his own power, as he will himselfe, for the preservation of his own Nature" and is what enables each person in their natural state to have the right to do "any thing, which in his own Judgement, and Reason" is likeliest to serve that end (2012, 198;

184   HOBBES'S TWO SCIENCES

1651, 64). Hobbes later portrays the "condition of Nature" as one in which "every man is Judge" (2012, 214; 1651, 70), and he describes each commonwealth as having this same sort of "absolute Libertie, to doe what it shall judge (that is to say, what that Man, or Assemblie that representeth it, shall judge) most conducing to their benefit" (2012, 332; 1651, 110).[15]

Leaving the state of nature requires, as the Second Law of Nature describes, that each person be willing to lay down their right to all things (2012, 200; 1651, 64). But so far this description seems to be entirely about *acts* related to things, objects that one person could hinder another from using or acquiring. Indeed, Hobbes defines laying down a right as when someone "*devest* himself of the *Liberty*, of hindring another of the benefit of his own Right to the same" (2012, 200; 1651, 64). However, later in the Seventeenth Law of Nature, Hobbes is clear that private judgments about this or that act must also be surrendered if humans are to create peace in a commonwealth:

> And seeing every man is presumed to do all things in order to his own benefit, no man is a fit Arbitrator in his own cause: and if he were never so fit; yet Equity allowing to each party equall benefit, if one be admitted to be Judge, the other is to be admitted also; & so the controversie, that is, the cause of War, remains, against the Law of Nature. (2012, 238; 1651, 78)

Hobbes's worry in allowing each person to serve as a judge in matters that relate to their own benefit is not only that they will be partial to themselves in this or that matter but—much more significantly—that this ability to judge privately would need to be distributed and each person would judge privately for themselves. Given such a general scenario where each person is "admitted to be Judge," there would be "controversie" that would lead to war.

The Seventeenth Law of Nature thus not only outlaws partiality in judgment but, because of the requirement for equity in benefits among subjects in the commonwealth, it outlaws all private judgments for all subjects, whether concerning religious matters or not. In *Leviathan* 18, Hobbes makes clear that the authorization of the sovereign concerns not just actions performed but "all the Actions and Judgements, of that Man, or Assembly of men, in the same manner, as if they were his own [. . .]" (2012, 264; 1651, 88). Furthermore, the rights of the sovereign include "the Soveraign Power, to be Judge, or constitute all Judges of Opinions and Doctrines, as a thing necessary to Peace; therby to prevent Discord and Civill Warre" (2012, 272; 1651, 91) as well as the "Right of Judicature; that is to say, of hearing and deciding all Controversies, which may arise concerning Law, either Civill, or Naturall, or concerning Fact" (2012, 274; 1651, 91). In *Leviathan* 20,

---

[15] This comparison is unsurprising given the analogy Hobbes makes in *Leviathan* 13 between individuals in the state of nature and the sovereigns of commonwealths (2012, 196; 1651, 63).

EXPERIENCE IN HOBBESIAN CIVIL PHILOSOPHY   185

Hobbes reiterates that the sovereign is the "Judge of what is necessary for Peace; and Judge of Doctrines: He is Sole Legislator; and Supreme Judge of Controversies" (2012, 306; 1651, 102). Without the powers of judgment being invested solely in a sovereign, each subject would serve as private judge and a state of war would return.

It is thus unsurprising that the presence of private judgment is one of the "*Diseases* of a Common-wealth" that Hobbes describes in *Leviathan* 29. He notes that the

> [. . .] sedition doctrine [. . .] *That every private man is Judge of Good and Evil actions* [. . .] is true in the condition of meer Nature, where there are no Civill Lawes; and also under Civill Government, in such cases as are not determined by the Law. But otherwise, it is manifest, that the measure of Good and Evil actions, is the Civill Law; and the Judge the Legislator, who is always Representative of the Commonwealth. (2012, 502; 1651, 168)

Although in a stable commonwealth this right to judge is held locally by the "Mortal God" who is sovereign, it is ultimately held by God alone. Indeed, Hobbes's strange exegesis of Original Sin in Genesis locates the Fall of Adam and Eve not in their disobedience to God's command regarding the fruit of the Tree but rather in their usurping God's office of Judicature: "Whereupon having both eaten, they did indeed take upon them Gods office, which is Judicature of Good and Evill; but acquired no new ability to distinguish between them aright" (2012, 318; 1651, 106).

Adam and Even underwent no material change after consuming the fruit; their sin originated in judging their nakedness as "uncomely" and in feeling ashamed they "tacitely censure[d] God himselfe" (2012, 318; 1651, 106).[16] Hobbes's exegesis has the following logic. Given that God created Adam and Even naked, prior to that act God must have judged that their nakedness was good (exemplified by God's statements about the goodness of creation following each day). However, in feeling shame, Adam and Eve judged nakedness to be bad and, in doing so, took on the office of judge themselves without the right to do so. This linkage between judgment and action is consistent with the statements earlier of the rights that are vested in the sovereign.[17]

---

[16] Hobbes categorizes shame as a species of the simple passion grief in *Leviathan* 6. Grief is a form of displeasure that results from "Expectation of consequences" (2012, 84; 1651, 25), and shame occurs upon the "discovery of some defect of ability" (2012, 90; 1651, 27). Although Hobbes holds that shame involves bodily components, such as "blushing," in the 1651 edition he uses terms with a cognitive valence like 'discovery' and 'apprehension'. The Latin emphasizes the cognitive (judgmental) aspect of shame by adding that it involves thinking: "*Griefe*, when we think [*putamus*] that something we have done is unseemly" (2012, 91).

[17] The 1651 edition suggests "that the Commands of them that have the right to command, are not by their Subjects to be censured, nor disputed" as an allegorical interpretation of God's question "Hast thou eaten?" (2012, 320; 1651, 106). The Latin 1668 edition alters this by making it about the "deeds" of sovereigns rather than their commands; despite this slight difference, for both versions *judgments* precede commands and deeds and so censuring either implies censuring the judgment that came before it.

186 HOBBES'S TWO SCIENCES

Returning to the explanation of the Civil War in *Behemoth*, the "facts" that speaker A introduces relate to the widespread availability of translations of Scripture in vernacular languages. This availability was followed by individual persons taking themselves to be their own "Judge of Religion" (2010, 135). These facts alone would not explain *why* they were part of the cause of the Civil War. However, as argued in chapter 4, since Hobbes takes the Laws of Nature *jointly* to be the cause of PEACE, missing any of those acts that satisfy one of the Laws would result in a return to WAR. The *fact* that group 1 (Presbyterians) and group 2 (Papists) had differing private judgments regarding the interpretation of Scripture implies a scenario in violation of Law Seventeen, and in such a situation Hobbes is clear that "the cause of War" (2012, 238; 1651, 78) remains. In short, the absence of the actions related to a single Law is sufficient to explain the conflict that arose, just as a violation of any one of the instructions for constructing SQUARE would provide the reason why something other than a SQUARE was made.

6.2.2 The Third Law of Nature and the "Sum" of the Laws in *Behemoth*

As A's use of the Seventeenth Law involved explaining the background conditions that were part of the entire cause of the Civil War, the Third Law of Nature and the "sum" of the Laws figure in the explanation of other events that led up to the Civil War. The invocation of these Laws as causes occurs within a discussion of a refusal to pay "shipmoney", which was a tax imposed outside of Parliamentary approval, to support the building and maintenance of ships commissioned by the King. Typically, the tax was assessed only on maritime counties, but Charles I attempted to use it to tax non-maritime counties during the 1630s.[18] Speaker A, as Hobbes's mouthpiece, argues that this taxation was not prohibited by the *Magna Carta*, interpreting that work as "[. . .] made not to exempt any many from payments to the publick, but for securing of euery man from such as abused the Kings Power by surreptitious obtaining the Kings warrants to the oppressing of those against whom he had any suit in Law" (2010, 157).

Members of Parliament objected to the shipmoney tax, A argues, because certain "rebellious" Members interpreted the *Magna Carta* incorrectly to serve their own ends. This interpretation was sufficient for the "understanding of the rest" of Parliament, so they "let it passe" (2010, 157). B worries that A's assessment reflects badly on the subjects who "chose [the Members of Parliament] for the wisest of the Land" (2010, 157). In other words, if subjects had chosen Members of Parliament who were truly wise, then those who weren't rebellious would not have voted in

---

[18] For details regarding this tax, see Seaward's General Introduction in the Clarendon *Behemoth* (Hobbes 2010, 34–35).

favor of the wrongful interpretation. B suggests this happened because "common people" are ignorant of their duties, "neuer meditating anything but their particular interest," and when deciding about that which is outside of their immediately apparent interests, they simply follow "Preachers" or the "most potent of the Gentlement amongst them" (2010, 158). A's response implies that while this may describe many situations like those leading up to the Civil War, it need not be this way if subjects were instructed properly: "Why may not men be taught their duty, that is the Science of Just and Vnjust" (2010, 158)?

At first glance, A's reply that the "Rules of *Just* and *Vnjust* sufficiently demonstrated, and from Principles euident to the meanest capacity, haue not been wanting [. . .]" (2010, 158–159) may seem simply like Hobbes's thinly-veiled attempt to repeat the "sum" that he offers in *Leviathan* 15. As discussed in chapter 4, Hobbes asserts that all of the Laws of Nature can be "contracted into one easie sum, intelligible, even to the meanest capacity; and that is, *Do not do to another, which thou wouldest not have done to thy selfe*" (Hobbes 2012, 240; 1651, 79). Speaker A is clearly echoing the account in *Leviathan* 15 with the language of "meanest capacity" as well as A's implied agreement with B that many individuals are simply "too busie in getting food, and the rest too negligent to understand" the potentially "too subtle deduction" of the Laws of Nature that *Leviathan* provides (Hobbes 2012, 240; 1651, 79).

However, rather than understanding speaker A's allusion to the "sum" as mere regurgitation of the material from *Leviathan*, I wish to draw attention to what follows in *Behemoth*. Speaker A draws attention to the ignorance of the many subjects who cannot read and the fact that even if some can read, either they "haue no leisure" or are interested in pursuing their "priuate businesses or pleasures" (2010, 159). Recall Hobbes's requirements for *scientia* outlined in chapter 3. Hobbes holds that for the development of *scientia* there must be leisure, curiosity, and proper method (2012, 322; 1651, 107). Speaker A clearly agrees with these criteria and is attributing the fact that the subjects in question were unaware of their duties either to a lack of leisure or, for those with leisure, to a lack of curiosity, i.e., the passion aimed at *scientia*: "*Desire* to know [1668: *sciendi*] why, and how" (2012, 86; 1651, 26).

Since they lack leisure combined with curiosity, subjects would be unable to arrive at *scientia* on their own. What is needed is for someone else, whom these subjects trust, to teach them the "sum". Hobbes through A suggests that the place where such individuals "should euer learne their duty" would be from the Pulpit, but this is precisely where such individuals "learned their disobedience" (2010, 159). Speaker A's indictment of Preachers ultimately rests on an unsurprising source—universities—which A claims "haue been to this Nation, as the wooden horse to the Troianes" (2010, 159). The role universities play in commonwealth maintenance and decay is a frequent topic in Hobbes's writings, but in this context of explaining the cause of the Civil War they play a role for disseminating knowledge related to subjects' duties to Preachers, who in turn deliver that knowledge to

188   HOBBES'S TWO SCIENCES

subjects during religious services or in their other capacities. In other words, the failure of universities to educate ministers is the first cause in a series of events that can lead to subjects being ignorant of their duties.

After a digression regarding the advent of universities and the connection they have had to educating preachers (2010, 159–161), Speaker A argues that Aristotelian philosophy from the universities was "made an ingredient to Religion" because it seemed to make certain "absurd Articles" of the Roman Church intelligible. The reason these articles took hold, A suggests, is because they served the ends of those in the Roman Church (discussed in section 6.3). Insofar as the universities have taught "Aristotles Ethicks," Speaker A suggests that there has been no harm to those in the Roman Church but also no help to subjects. A claims that while there have been many disagreements about what virtue and vice are, there has been "no knowledge of what they are, nor any method of attaining Vertue, nor auoiding Vice" (2010, 164).

Speaker A next distinguishes the virtue of subjects from that of sovereigns. The virtue of subjects, A argues, is "comprehended wholly in obedience to the Laws of the Common wealth" (2010, 165). A appeals to the Laws of Nature to support this claim:

> To obey the Laws is Justice and Equity, which is the Law of Nature, and consequently is Ciuill Law in all Nations of the world. And nothing is Iniustice or Iniquity otherwise then it is against the Law. (2010, 165)

Here Hobbes, through Speaker A's mouth, seems to be referring to a collection of the Laws of Nature. Indeed, Hobbes sometimes refers to the Laws of Nature generally by referring to single terms that pick them out ("Justice and Equity"). For example, he refers to the complete set of the Laws of Nature in this way when concluding *Leviathan* 15: "[. . .] consequently all men agree on this, that Peace is Good, and therefore also the way, or means of Peace, which (as I have shewed before) are *Justice, Gratitude, Modesty, Equity, Mercy,* & the rest of the Laws of Nature, are good; that is to say, *Morall Vertues;* and their contrarie *Vices,* Evill" (2012, 242; 1651, 50). In the context of *Behemoth*, with the phrase "Justice and Equity" Speaker A seems to be referring to at least the Third, Eleventh, Twelfth, and Thirteenth Laws. I will discuss each of these briefly to connect them to the context of A's explanation in the First Dialogue.

That speaker A has in mind the Third Law of Nature, which prescribes "*That men performe their Covenants made,*" is implied by A's reference to the nature of justice being obeying the laws. Hobbes defines injustice in *Leviathan* 15 as simply not performing a covenant and states that "the nature of Justice, consisteth in keeping valid Covenants" (2012, 220; 1651, 72). This connection to the Third Law is made stronger by Speaker A's reference immediately following the extended quotation above to persons who fit the description of the "Foole" in *Leviathan* 15. Speaker

A describes such individuals as "priuate men" who wrongly see it as "Prudence" to "enrich themselves; yet craftily withhold from the publick, or defraud it" from taxes or other funds that are legally imposed" (2010, 165). Just like the Foole in *Leviathan* 15, who thinks that he has reason to break the covenant, "cannot be received into any Society, that unite themselves for Peace and Defence" unless those in that society make a mistake (2012, 224; 1651, 73), speaker A's "priuate men" wrongly think it is prudent to defraud fellow parties to the covenant and so must lack "knowledge of what is necessary for their own defence" (2010, 165). Both individuals—the Fool in *Leviathan* and the one who thinks cheating is prudent in *Behemoth*—think that they are benefiting themselves but in the end work against their own interests. The Foole is doomed, as Hobbes argues in *Leviathan* 15, either to be cast out of society or accepted only because their fellow subjects are ignorant of what is good for themselves.

Speaker A's reference to equity suggests that leading up to the Civil War the subjects were not being taught the Laws of Nature that relate specifically to equity, which Hobbes understands as distributive justice (2012, 230; 1651, 75). Laws clearly grounded in equity are the Eleventh (judges should be impartial), Twelfth (things should be equally divided or, if not possible, equally used), and Thirteenth (lot should be used to determine rights to things that can be neither divided nor used commonly) Laws of Nature (2012, 236; 1651, 77).

I have suggested that these references by speaker A in *Behemoth* to the Laws of Nature should not be taken simply as Hobbes trotting out the claims of *Leviathan* in a new context. Instead, seeing them as Hobbes borrowing the Laws as causal principles to explain *why* the Civil War happened makes sense of A's claim to be providing part of the "seeds" of the War. When an individual or group does not follow the Laws, PEACE will not be created and thus one would expect that disorder would result because there has been a return to WAR. The facts are established by what speaker A recounts, for example, related to the founding of the universities, the details related to various Aristotelian doctrines and university curricula, along with the topics of sermons from Preachers. These can be established by experience, but they could not provide an explanation for *why* these events coming together at the same time resulted in the Civil War.

Only speaker A's invocation of the Laws provides the reason why the War resulted, for any situation in which one or more of the Laws is not being followed would result in a circumstance of WAR. All that was required to avoid such a state was for Preachers or other respected leaders to learn the Laws of Nature, such as those related to Justice and Equity, at universities and then to teach subjects the "sum" so that even those with the "meanest" intellect could understand their duties. In the absence of such knowledge, whether of all of the Laws singly or simply the "sum", Hobbes argues that mere personal experience (prudence) will be used, for better or for worse: "Ignorance of the causes, and originall constitution of Right, Equity, Law, and Justice, disposeth a man to make Custome and Example the

190   HOBBES'S TWO SCIENCES

rule of his actions; in such manner, as to think that Unjust which it hath been the custome to punish; and that Just, of the impunity and approbation whereof they can produce an Example [...]" (2012, 158; 1651, 50).

6.3 The "Naturall Cause of Religion" and the Kingdom of Darkness

6.3.1 Humans' Natural State of Ignorance, Curiosity, and Anxiety

A primary aim of Part IV "Of the Kingdom of Darkeness" of *Leviathan* is to explain how the liberty of subjects came to be bound by the Roman Church. Showing these steps—what Hobbes identifies as "knots"—involves both recounting historical events (the 'that') and appealing to causal principles from the science of civil philosophy (the 'why'). Hobbes identifies the joining of these "knots" as "the whole *Synthesis* and *Construction* of Pontificall Power" in *Leviathan* 47 (2012, 1114; 1651, 385). He identifies three: first, presbyters, who coordinated with each other through assemblies, convinced believers that they were obliged to obey their doctrine; second, as groups of presbyters increased, they became governed by bishops; and third, the bishop of Rome ascended to rule over all other bishops by the wills both of the other bishops as well of the emperors when their power was waning.

In this section, I argue that Hobbes's explanation of Papal bondage is structurally analogous to Hobbes's explanations in natural philosophy where he mixes both the 'that' and the 'why'. Rather than seeing Hobbes as "recycling his material" from his other works, as Springborg (1994, 569–570) suggests, I argue that in *Leviathan* Part IV Hobbes borrows in the same way that he does when citing causal principles from geometry or principles from first philosophy in *De corpore* Part IV. In the case of explaining the bondage of subjects to the Pope, Hobbes borrows the 'why' from *Leviathan* 12, where he explicates the starting point and provides the "naturall cause of religion." I first discuss this causal explanation and then show how it functions as a borrowed cause in Part IV to explain the ascendancy of the Bishop of Rome.

Since in chapter 4 I located the genesis of the science of civil philosophy in the thought experiment of *Leviathan* 13, it might seem strange to claim here that Hobbes borrows a causal principle from the science of civil philosophy from *Leviathan* 12 to explain Papal bondage. However, it is unproblematic that the material in *Leviathan* 12 precedes the state of nature thought experiment because, as I discuss, Hobbes begins both accounts in the same root—the equality of human bodies (EQUAL). As argued in chapter 4, the state of nature thought experiment begins by explicating EQUAL: human bodies (HUMAN) considered as apart from civil relationships have the property EQUAL, understood as the ability for the weakest to kill the strongest (Hobbes 2012, 188; 1651, 60). In that context,

Hobbes sees even greater equality evident in the faculties of mind because, in their natural state, humans with equal time will develop equal experience and thus acquire equal levels of prudence (Hobbes 2012, 188; 1651, 60–61). The account he provides next moves to HOPE and traces how the state of nature is one in which all parties have a continual fear of death.

The explanation Hobbes provides for the "naturall cause of religion" ultimately appeals to ANXIETY, and prudence is the root of AXIETY. Since prudence is grounded in EQUAL, as I discuss below, the account of the natural cause of religion begins at the same starting point as the thought experiment in *Leviathan* 13—the reasoning simply follows a different trajectory. As discussed in chapter 2, Hobbes holds that humans and non-human animals alike share in prudence; indeed, some non-human animals are more prudent than some humans (2012, 44; 1651, 10–11). But humans are unique in that in their natural state they are moved by the passion of curiosity. This allows humans to act as makers of things, like geometrical figures and laws for commonwealths (the path toward making PEACE), but in *Leviathan* 12 Hobbes suggests that curiosity also naturally leads down a different path—to the development of religion. Considering HUMAN in this state shows that HUMAN contains CURIOSITY, or as Hobbes puts it, "[. . .] it is peculiar to the nature of Man, to be inquisitive into the causes of events they see, some more, some less; but all men so much, as to be curious in the search of the causes of their own good and evil fortune" (2012, 164; 1651, 52).

Hobbes is not concerned in *Leviathan* 12 with delineating "religion" from "superstition," since he distinguishes the two simply by holding that the former is "publiquely allowed" (2012, 86; 1651, 26). Instead, his aim is to provide a causal account of how religious belief, whether "superstition" or not, arises in human bodies (HUMAN). Thus, his explanation of the natural cause of "religion" is no different in status from his explanation of how, say, DIFFIDENCE arises in the state of nature (both are cases of Making Type 3 discussed in chapter 3).

In addition to being moved by the passion of CURIOSITY, Hobbes suggests in *Leviathan* 12 that humans naturally trace events back to their beginnings and hold that there must be some cause to determine every effect. As he puts it, "upon the sight of any thing that hath a Beginning, to thing also it had a cause, which determined the same to begin, then when it did, rather than sooner or later" (2012, 164; 1651, 52). These two features of humans (HUMAN explicated as containing CURIOSITY + ATTRIBUTION OF CAUSES) in their natural state combine to *produce* ANXIETY that one will not be able to acquire the things one wants given the potential for one's desires to be thwarted.

These considerations of HUMAN in *Leviathan* 12 are not explicitly within the form of a thought experiment, but they fit the mold of an explication (discussed in chapter 4, section 4.1); they result from considering what is contained in HUMAN, and Hobbes intends these considerations to be general in scope like we would expect were they in the form of a thought experiment. Hobbes states that "every man" has ANXIETY but some are "over provident" and have it in greater measure.

192 HOBBES'S TWO SCIENCES

Furthermore, Hobbes describes the state in which ANXIETY arises in the same way as humans' natural state: just as humans' natural state is characterized by "continuall feare, and danger of violent death" and the humans in it live a life that is "solitary, poore, nasty, brutish, and short" (2012, 192; 1651, 62), humans with ANXIETY are plagued by "perpetuall feare," are constantly "gnawed on by feare of death, poverty, or other calamity," and have "no repose, nor pause of anxiety, but in sleep" (2012, 166; 1651, 52).

Hobbes makes clear that ANXIETY is traceable to human prudence, and he compares it to Prometheus's torment by linking it to the etymology of his name in Greek—"thinking ahead"—which he identifies as *The prudent man* (2012, 166; 1651, 52). Given the praise Hobbes offers for prudence earlier in *Leviathan*, it may seem surprising to see him compare the over-prudent individual with Prometheus's perpetual suffering caused by an eagle eating his liver during the day followed by the regeneration of his liver each night. The life of the anxious individual in their natural state is one which is consumed by thoughts of the future and the perils that might await them, a constant "feare of death, poverty, or other calamity" such that the individual is unable to sleep (2012, 166; 1651, 52). This fear arises because such an individual does not know the causes that are responsible for many of the events of their lives, because they do not see those causes, and so they attribute the events of their life to something unobservable like fate or to some invisible agent like a deity.

Tying in this connection to prudence, the causal story so far goes as follows: individuals in their natural state are EQUAL and thus develop PRUDENCE when they undergo the same experiences. While PRUDENCE brings benefits, in their natural state those with PRUDENCE find themselves in continual fear that they will be unable to keep themselves from harm and satisfy their desires. CURIOSITY + ATTRIBUTION OF CAUSES + IGNORANCE OF CAUSES + PRUDENCE causes ANXIETY. Humans with ANXIETY look to find relief, but in many cases, there is no observable cause to help them understand why events occurred or how to encourage them or prevent them from occurring in the future. As a result, they invent

> [...] some *Power*, or Agent *Invisible*: In which sense perhaps it was, that some of the old Poets said, that the Gods were at first created by humane Feare: which spoken of the Gods, (that is to say, of the many Gods of the Gentiles) is very true. (2012, 166; 1651, 52–53)

Hobbes is careful to differentiate the role fear plays in explaining religious beliefs of the Gentiles and their polytheistic systems from the belief in the "one God Eternal, Infinite, and Omnipotent" (2012, 166; 1651, 53). Unlike the case of polytheistic systems, Hobbes suggests that holding that there is a First Mover is not based in fear but simply in the human desire to understand the causes of natural things and, through a series of inferences, arriving at the belief that there must be some un-caused cause from which all motion originated.

In addition to the positing of invisible agents responsible for events, in *Leviathan* 12 Hobbes proposes three additional "seeds" of religion: an ignorance of second causes, a devotion to or worship of the things that are feared, and the belief that things that happen fortuitously are signs of future things to come (2012, 170, 1651, 54). Hobbes explicates ignorance of second causes as the condition of *merely* prudent individuals:

> [...] men that know not what it is that we call *causing*, (that is, almost all men) have no other rule to guesse by, but by observing, and remembering what they have seen to precede the like effect at some other time, or times before, without seeing between the antecedent and subsequent Event, any dependence or connexion at all: And therefore from like things past, they expect the like things to come; and hope for good or evill luck, superstitiously, from things that have no part at all in cause it. (2012, 168; 1651, 53)

Individuals—as Hobbes says, this characterizes "almost all men"—in this condition express their reverence to the invisible agents that they posit by giving thanks, giving gifts, acting in certain ways, and so on, just like they would to a visible cause (2012, 170, 1651, 54). These individuals take fortuitous or chance happenings to be the way that these invisible agents "declare" to them what will happen in the future. Adding to the earlier causal account, Hobbes has now shown that DEVOTION (worship) and PREDICTION of the future follow from ANXIETY. Hobbes concludes *Leviathan* 12 by arguing that since religion has natural causes it can never be fully eliminated. Instead, a religion can be modified or changed for another; he identifies this as the "resolution of [a religion] into its first seeds" (2012, 180; 1651, 58).[19]

Hobbes's account in *Leviathan* 12 thus begins with an explication of EQUAL humans and shows how such individuals come to have religious beliefs. This "naturall cause" explains beliefs, both cases of religion as such, i.e., that which is permitted, and cases of superstition. This causal explanation serves as the 'why' for the historical explanation of how the liberty of Christians was bound by the actions of the presbyters and, eventually, by the ascension of the Bishop of Rome. The next subsection turns to the role this cause plays in that explanation.

---

[19] Stauffer (2018, 111–112) suggests that in this concluding section of *Leviathan* 12 Hobbes might have implied that it was possible for religion to be fully eradicated. Stauffer draws support for this claim from the Latin *Leviathan* where Hobbes adds the caveat "if suitable cultivators emerge" when describing the possibility of another religion arising after one is dissolved. I do not address this possibility because my interest is only in the role that the causal explanation plays in Hobbes's later historiographical account.

194 HOBBES'S TWO SCIENCES

6.3.2 "The Night of our naturall Ignorance"

A casual reading of *Leviathan* 44 would seem to make Hobbes's explanation for the ascendancy of and domination by the Bishop of Rome *ad hoc*. He begins the chapter, which opens Part IV, by positing "four causes of spiritual darkness" (2012, 958; 1651, 334):

1. The misinterpretation of scripture;
2. Deceptions involving belief in demons;
3. Mixing scriptural interpretation with philosophy, in particular Aristotelianisms; and
4. Mixing scriptural interpretation with false and uncertain historical claims.

Considered alone, these four "causes" seem merely speculative and perhaps specially crafted to fit Hobbes's rhetorical ends against the Roman Church. Furthermore, given his restriction that *scientia* involves knowing *actual* causes, which knowledge is available only to makers, he clearly cannot mean that these "causes" are of equal epistemic standing with those of geometry or civil philosophy. As a result, it might be tempting to think that his use of 'cause' should not be taken seriously. However, if the mixed-mathematical model of explanation is Hobbes's inspiration for historiographical explanations, as I am arguing, this claim about "causes" should be understood as on the same epistemic level as his explanation of the "cause" of sense in *De corpore* 25, discussed in chapter 5, which incorporates both the 'why' from geometry and the 'that' from observations.

Hobbes's introduction of these four causes of spiritual darkness begins with an immediate connection to *Leviathan* 12, where IGNORANCE OF CAUSES is part of the explanation of religious belief: "The Enemy has been here in the Night of our naturall Ignorance, and sown the tares of Spirituall Errors" (2012, 958; 1651, 334). The Latin emphasizes that this was a "dark night [1668: *obscurâ nocte*]." Hobbes's readers would have been drawn to think about Jesus's parable in Matthew 13:24–30 where the Sower's enemy planted weeds ("tares") in the field while he slept. In what remains of this section, I briefly show that two of these four causes of spiritual darkness (Cause 1 and Cause 2 above) each draw upon the causal explanation in *Leviathan* 12. Hobbes's aim to show the three "knots" that bound Christian believers, culminating in *Leviathan* 47, combines facts related to ecclesiastical history (the 'that') with causal principles from civil philosophy (the 'why'). My focus will be on points where Hobbes is making use of civil philosophy, but just like Hobbes's natural philosophy is not entirely "mixed," not all parts of this historiographical explanation are mixed.

In the remainder of *Leviathan* 44, Hobbes describes misinterpretations of Scriptural texts that he claims were one of the causes of spiritual darkness (Cause 1). Some of the misinterpretations exploit the causal explanation from *Leviathan*

12, but not all of them do this. For example, Hobbes's claim that the Roman Church has erroneously asserted itself as the "Kingdome of God" on Earth does not rely upon the earlier causal explanation. Hobbes's criticism of that claim relies on textual evidence that there had not been a Kingdom of God since the Jewish people requested Saul to be appointed as king (2012, 960; 1651, 335) and thus the Roman Church could not rightfully claim that title.[20] In saying that other misinterpretations to be discussed exploit the causal explanation, I mean the following: the fact that some interpretation or other happened is a matter of historical record, but the reason *why* that interpretation was successful in convincing Christian believers to accept their eventual bondage relies upon using a cause from *Leviathan* 12.

I first discuss two of Hobbes's examples related to Cause 1 (dogmas related to the Eucharist and to baptism). Hobbes claims that spiritual darkness resulted from the "turning of Consecration into Conjuration," the latter of which was done by nefarious individuals who exploited ANXIETY by appealing to invisible agents and their supposed powers as well as by using expressions of reverence such as "Petitions, Thanks, [and] [. . .] premeditated Words" (2012, 170; 1651, 54). When described in scriptural texts, Hobbes states that 'consecration' simply meant to separate something "from common use" (2012, 966; 1651, 337). Many things were consecrated in scriptural texts (Hobbes cites his discussion from *Leviathan* 35 again). The point of such an act, Hobbes emphasizes, was not to change the thing that was consecrated but rather only "the use of it." In other words, consecrating anything, such as a goblet or other serving vessel, only set it apart for certain functions.

Priests and presbyters, however, preyed upon the natural ignorance of Christian believers and suggested to them that the words of consecration changed the nature of the objects themselves, using it as no other than a "*Conjuration* or *Incantation*, whereby they would have men to believe an alternation of Nature that is not, contrary to the testimony of mans Sight, and of all the rest of his Senses" (2012, 966; 1651, 337). Importantly, Hobbes does not suggest that the conjurations, such as the Roman Church's claim that Christ's words "This is my Body" and "This is my Blood" implied transubstantiation (examples of "premeditated words"), were forced upon the people with threats. Instead, the reason *why* this misinterpretation took hold was because it was introduced during a time when the ignorance of believers was so great—as Hobbes says, "[. . .] when the Power of the Popes was at the Highest, and the Darknessse of the time grown so great" (2012, 970; 1651, 338)—that they were not sufficiently discerning to see that the doctrine of transubstantiation conflicted with evidence from their senses. That doctrine, Hobbes notes, is on even worse epistemic footing than the claims of sorcerers whose

---

[20] As part of this textual evidence from Scripture, Hobbes cites what he has "proved" in *Leviathan* 35 (2012, 634–640; 1651, 216–218).

196 HOBBES'S TWO SCIENCES

incantations at least change the sensible appearances of objects, such as Hobbes's example of Pharaoh's conjurers who changed rods into snakes. There are two parts to this example of Cause 1: first, the historical fact *that* transubstantiation and other doctrines were introduced; and second, the reason *why* those doctrines were successful, which lies in the causal account of natural human ignorance, the tendency to accept invisible agents, and the use of premeditated words. If human ignorance did not lead (according to the *Leviathan* 12 causal account) to such a result, then invoking *that* such events transpired would fail to count as a causal explanation.

The second example of Cause 1 makes the same point as the first but about a different sacrament: rather than simply consecrating the elements to be used in the ceremony of baptism, each is treated as if it were changed substantially through an exorcism, and likewise the infant is subjected to "many Charms" of incantations (2012, 970; 1651, 338–339). Throughout the ceremony, priests appeal to invisible agents and their powers and utter "premeditated words"; both features of the ritual are present in the causal account from *Leviathan* 12. Hobbes's claim is that priests exploited natural ANXIETY by using these rituals and this explains why ignorant people accepted bondage.

Deceptions involving belief in demons—Cause 2—are the subject of *Leviathan* 45. The chapter opens with a repetition of Hobbes's explanation of perception by pressure, presented originally in *Leviathan* 1 and 2 and in *De corpore* 25 (the 1668 Latin refers the reader to those chapters rather than repeating the account). This repetition is more than mere recycling of material, but it is not a borrowing of a causal principle (since we do not know the *actual* cause of sensation because it is within the domain of natural philosophy, only a possible cause). Instead, I suggest that Hobbes repeats this material to expand upon his pronouncement that humans are naturally ignorant in a way that leads to ANXIETY and ultimately to religion. Daemonology relies on a particular form of ignorance—ignorance about the nature of sensation:

> This nature of Sight having never been discovered [1668: *ignorarent*] by the ancient pretenders to Natural Knowledge; much lesse by those that consider not things so remote (as that Knowledge is) from their present use; it was heard for men to conceive of those Images in the Fancy, and in the Sense, otherwise, than of things really without us: Which some (because they vanish away, they know not whither, nor how,) will have to be absolutely Incorporeall, that is to say Immateriall, or Formes without Matter; Colour and Figure, without any coloured or figured Body [...]. (2012, 1012; 1651, 352)

Hobbes's diagnosis goes as follows: Ignorance of the nature of sensation results in thinking that phantasms that appear and disappear without an intelligible and observable cause must be the result of some immaterial agent. He ridicules this inference since it would be equivalent to saying that someone "saw his own Ghost in a Looking-Glasse, or the Ghosts of the Stars in a River" (2012, 1014; 1651, 352). Since humans naturally have a disposition to hold that an effect has a cause, and since they cannot observe the cause, given the causal account of *Leviathan* 12, Hobbes can conclude that they were inclined to attribute such phantasms to the work of invisible agents who had "an unlimited power to doe them good, or harme" (2012, 1014; 1651, 352–353).

This natural tendency of ignorant humans was exploited by many, both "Governours of Heathen Common-wealths" and the Roman Church. Hobbes offers some scriptural evidence against making the inference that incorporeal agents are responsible for such phantasms in the sections that follow, claiming "I have not yet observed any place in Scripture [...] that any man was ever possessed with any other Corporeal Spirit, but that of his owne" (2012, 1018; 1651, 354), and furthermore argues that there are no passages that explicitly describe incorporeal agents. Any speculation about incorporeal agents, he concludes, is just that—mere speculation about matters that are "more curious, than necessary for a Christian mans Salvation" (2012, 1020; 1651, 355). Nevertheless, the reason *why* the Roman Church's use of such speculation, which is part of the cause of the binding of believers, was successful lies in the ignorance of humans—because they were ignorant of the cause of sensation, they were susceptible to believing in demons and other invisible agents with unlimited power to help them or harm them.

Returning to the sketch Hobbes provides of the "synthesis" of Christian bondage by the Roman Church, the need for these causal principles borrowed from *Leviathan* 12 is evident. The "knots," to which Hobbes appeals to explain how the "web" encircled Christians, contain the causes described already. The first "knot"—when presbyters decided together what they should teach and that those who did not follow their commands would be excommunicated—succeeded because those "doctrines" exploited the tendencies of naturally ignorant humans. The same follows for the remaining two knots, which culminated in the Bishop of Rome assuming control over the other bishops. Hobbes's reliance upon causes is clear from his "resolution" of the knots immediately following the "synthesis". There Hobbes traces the breaking of first knot to the "Power of the Pope" being dissolved by Queen Elizabeth and the breaking of the second knot to Presbyterians "putting down the Episcopacy" (2012, 1114–1116; 1651, 385). The third knot was dissolved by the removal of the power of the Presbyterians[21]: "[...] so we are reduced to the Independency of the Primitive Christians to follow Paul, or Cephas, or Apollos" (2012, 1116; 1651, 385). In sum, the reason *why* Christian believers found themselves in bondage to the Roman Church is because the priests and ultimately the Pope exploited the natural ignorance of subjects and played upon

---

[21] Malcolm suggests this refers to the "Act for the Repeal of Several Clauses in Statutes imposing Penalties for not Coming to Church" in 1650 (see Hobbes 2012, 1116, fn. aa).

198 HOBBES'S TWO SCIENCES

their ANXIETY. This would not have counted as an explanation were it not for this linkage to the causal account in civil philosophy.

6.4 Two Potential Worries for Hobbesian Historiography

Two apparent differences between Hobbesian natural philosophy and Hobbesian historiography may seem to threaten the structural connection that I have been making between them. The present section will discuss each briefly and argue that these differences are a matter of content and not form. In other words, the structural analogy between historiography and natural philosophy still holds, even if the content of one may differ from the other in some ways. The first difference I discuss is that many of Hobbes's explanations in historiography are negative explanations (e.g., why PEACE was not preserved), where Hobbes explains why some result did not happen, but the explanations in natural philosophy are positive and treat why some event did happen. The second difference is that events explained in historiography appear to be one-time events and not repeated phenomena like those in natural philosophy.

The first apparent difference between historiography and natural philosophy can be dissolved by returning to Hobbes's understand of causation in terms of the "entire cause" (*causa integra*). This feature of Hobbes's philosophy was discussed in section 6.2 and earlier in chapters 3 and 4. Hobbes holds that a cause is "the aggregate of all the accidents both of the agents how many soever they be, and of the patient, put together" (OL I.107–108; EW I.121–122). As a result, Hobbes understood causes to be simultaneous to their effects as well as necessary for their production. This view helps make sense of using causes to explain both the *positive* presence of some feature as well as its *absence*. Since PEACE is caused necessarily and simultaneously when the actions described by all the Laws of Nature are performed all at once, the absence of any one of the actions specified by the Laws explains both the presence or absence of PEACE (and thus the presence or absence of WAR).

The second apparent difference relates to the *explananda* of historiography and natural philosophy. While natural philosophy seems to explain repeatable phenomena, like the explanation for sensation in *De corpore* XXV that is repeated many times over throughout the lives of every sentient being, historiography explains events that happen only a single time and never seem to be repeated. There will only ever be one unique Fall of Rome, for example, so it appears that natural philosophy deals with a fundamentally different kind of event—repeatable experiences in common among many perceivers.

Two replies can be made in response to this potential worry. First, Hobbes never makes repeatability a criterion for some event that is to be explained in natural philosophy. This is clear both in his own explanations as well as in his criticisms

EXPERIENCE IN HOBBESIAN CIVIL PHILOSOPHY   199

of others' explanations. For example, when explaining sense in *De corpore* XXV, Hobbes simply asserts the fact that the appearances of things continually change as the 'that': "it is proper in the first place to observe that our phantasms are not always the same, but new ones are constantly being created and old ones are disappearing, just as the organs of sense are turned now to one, now to another object" (OL I.317; EW I.389). While this does mention repeated experiences that sense perceivers share, it does not require that the reader engage in further experiences to be sure that this is repeatable.

Similarly, in Hobbes's criticisms of Boyle and others in *Dialogus Physicus*, speaker A never demands that the experimenters *repeat* their observations by running more experiments. Instead, A simply grants that the experimenters can take their observations as factual. For example, A notes that "[t]hey can be certain that the scale in which the bladder is, is more depressed than the other, their eyes bearing witness" (Hobbes 1985, 369; OL IV.261). Indeed, A explicitly denies any desire to controvert the experimenters' testimony:

> I do not wish to deny that the bladder, whether it be inflated by bellows or by blowing from the mouth, may be heavier than when the same bladder is not inflated, because of the greater quantity of atoms from the bellows or sooty corpuscles from the breath being blown in. (Hobbes 1985, 369; OL IV.261)

The criticism that A offers relates not to the repeatability of experiments/experiences, for A makes no demands to build more air-pumps so that others can recreate them, but instead A argues that the experimenters cannot know that the gravity of air is the *cause* of what they observe. The issue, according to A, is that the experimenters simply cannot know the cause of natural phenomena from observation alone, whether a single observation or many, *repeated* ones. As A puts it, "[. . .] they gather nothing sufficiently certain from the experiment made with the inflated bladder" (Hobbes 1985, 369; OL IV.261). The experimenters gain nothing "certain" because one can only hope to gain *cognitio* from sense experience/experiments, whether they are one-time events or a series of similar events.

Second, Hobbes's epistemology does not require that he think of each historical event as utterly unique. Instead, his epistemology, sketched in chapters 2 and 3, allows him to "consider as" historical events in certain ways depending upon his aim. There will always be features that set apart one event from another and make it unique, but this is no different from the way in which Hobbes's account of conceptions holds that every individual, distinct conception in the human mind is unique from all others. For example, the conception that I have of the pen in front of me right now will be distinct from the conception formed a second later, and two seconds later, and so on. Nevertheless, Hobbes's epistemology attempts to account for why knowers identify the same material object when referring to the 'pen' from moment to moment—they do so because they consider the material

200   HOBBES'S TWO SCIENCES

object in each conception as something that exists over time despite changing its properties. I suggest that Hobbes's nominalism can provide the same account regarding events of the historical past. Although each of them is utterly unique when considered simply as one-off events, when considering them in particular ways one can generalize on that basis. For example, it is intelligible for Hobbes to talk about the conflict in and fall of ancient democracies generally because in doing so he is considering certain features in common among different historical events while ignoring dissimilarities among them. The Hobbesian historiographer does just that.

6.5 Conclusion

This chapter has argued that Hobbesian historiography has an analogous structure to Hobbesian natural philosophy. In both disciplines, one uses observation to show 'that' something has occurred and then borrows the cause, the 'why', from the relevant science. In natural philosophy, a causal principle is borrowed from geometry; in historiography, a cause is borrowed from civil philosophy. Although Hobbes clearly excludes history from philosophy in his mature work *De corpore*, as he should do since the record books of history are not epistemically different from one's personal memories from sense, understanding a place for Hobbesian *historiography* enables us to make sense of Hobbes's assertion that history is "useful (no, indeed necessary [*necessarias*]) for philosophy" (1981, 189; OL I.9; EW I.10–11). This work of historiography is illustrated in Hobbes's explanations of various historical events, and the present chapter has considered two examples: the explanation in *Behemoth* of the Civil War in England and the explanation in *Leviathan* Part IV of the Pope's bondage of Christian subjects. In both works, Hobbes's reliance upon his claims in civil philosophy is not merely a recycling of that material but is instead the borrowing of a causal principle to show *why* events turned out the way that they did.

Works Cited

Primary Works

Aubrey, John. 1898. *Brief Lives*. Vol. I. Andrew Clark, ed. Oxford: Clarendon.

Barrow, Isaac. 1860 [1670]. *Lectiones Geometricae*. In William Whewell, ed., *The Mathematical Works of Isaac Barrow*, vol. I. Cambridge: Cambridge University Press.

Barrow, Isaac. 1860 [1685]. *Lectiones Mathematicae*. In William Whewell, ed., *The Mathematical Works of Isaac Barrow*, vol. I. Cambridge: Cambridge University Press.

Barrow, Isaac. 1734. *The Usefulness of Mathematical Learning explained and demonstrated:* Being Mathematical Lectures . . . , John Kirkby, trans. London.

Berkeley, George. 1949. *The Works of George Berkeley, Bishop of Cloyne*. Vol II. London: Thomas Nelson & Sons Ltd. Cited as *Works* with volume and number.

Boyle, Robert. 1682. *An Examen of Mr. T. Hobbs and his* Dialogus Physicus De Natura Aeris. London: Printed by M. Flesher, for Richard Davis.

Descartes, René. 1964–. *Oeuvres de Descartes*. Adam and Tannery, eds. Paris: J. Vrin. Cited as AT with volume and page number.

Descartes, René. 1985. *The Philosophical Writings of Descartes*, Vols. 1 and 2, J. Cottingham, R. Stoothof, and D. Murdoch, trans. Cambridge, MA: Cambridge University Press. Cited as CSM with volume and page number.

Descartes, René. 1991. *The Philosophical Writings of Descartes*. Vol. 3, The Correspondence, J. Cottingham, R. Stoothof, D. Murdoch, and A. Kenny, trans. Cambridge, MA: Cambridge University Press. Cited as CSM with volume and page number.

Descartes, René. 2001. *Discourse on Method, Optics, Geometry, and Meteorology*. Indianapolis: Hackett.

Hobbes, Thomas. 1640/1928. *The Elements of Law*, F. Tönnies, ed. Cambridge: Cambridge University Press. Cited as EL with part, chapter, and article number.

Hobbes, Thomas. 1650. *Humane Nature: Or, The fundamental Elements of Policie*. London: Printed by T. Newcomb.

Hobbes, Thomas. 1655. *De corpore*. London: Andrew Crooke.

Hobbes, Thomas. 1656a. *Six Lessons to the Professors of Mathematiques, one of Geometry, the other of Astronomy*. London: Andrew Crooke.

Hobbes, Thomas. 1656b. *Concerning Body*. London: R. & W. Leybourn, for Andrew Crooke.

Hobbes, Thomas. 1661. *Dialogus Physicus, sive De natura Aeris*. London: A. Crook.

Hobbes, Thomas. 1662. *Problemata Physica*. London: Andrew Crooke.

Hobbes, Thomas. 1666. *De Principiis et Ratiocinatione Geometrarum*. London: A Crooke.

Hobbes, Thomas. 1674. *Principia et Problemata alquiquot Geometrica*. London: Gulielmo Crook.

Hobbes, Thomas. 1678. *Decameron Physiologicum*. London: William Crook.

Hobbes, Thomas. 1682. *Seven Philosophical Problems, and Two Propositions of Geometry*. London: William Crook.

Hobbes, Thomas. 1839–1845. *The English works of Thomas Hobbes*, 11 volumes. Sir William Molesworth, ed. London: John Bohn. Cited as EW with volume and page number.

Hobbes, Thomas. 1839–1845. *Thomae Hobbes malmesburiensis opera philosophica*, 5 volumes. Gulielmi Molesworth, ed. London: John Bohn. Cited as OL with volume and page number.

Hobbes, Thomas. 1963. *Tractatus Opticus II*. Transcribed in Alessio, F., ed. "Thomas Hobbes: *Tractatus Opticus*." *Rivista Critica di Storia della Fiiosofia* 18: 147–228.

Hobbes, Thomas. 1973 [1642–1643]. *Critique du De Mundo de Thomas White*. Jean Jacquot and Harold Whitmore Jones, eds. Paris: Vrin.

202   WORKS CITED

Hobbes, Thomas. 1976 [1642–1643]. Thomas White's *De Mundo* Examined. Harold Whitmore Jones, trans. London: Bradford University Press.

Hobbes, Thomas. 1981. *Computatio sive Logica: Logic.* A.P. Martinich, trans. and commentary, Isabel C. Hungerland and George R. Vick, eds. New York: Abaris Books.

Hobbes, Thomas. 1983 [1646]. "*Thomas Hobbes's* A Minute or First Draught of the Optiques: *A Critical Edition.*" Elaine C. Stroud, ed. PhD diss., University of Wisconsin-Madison.

Hobbes, Thomas. 1985. *Dialogus Physicus, sive De natura Aeris.* Translation of the 1661 edition of Dialogus Physicus by Simon Schaffer in Shapin and Schaffer (1985).

Hobbes, Thomas. 1994a. *The Correspondence of Thomas Hobbes,* two volumes, Noel Malcolm, ed. Oxford: Clarendon Press.

Hobbes, Thomas. 1994b [1658]. *Man and Citizen*: De homine and De cive. Bernard Gert, ed. and trans. Indianapolis, IN: Hackett

Hobbes, Thomas. 1998. *Hobbes: On the Citizen.* Richard Tuck and Michael Silverthorne, eds. and trans. Cambridge: Cambridge University Press.

Hobbes, Thomas. 1999. *De corpore.* Critical Edition. Karl Schuhmann, ed. Paris: Vrin.

Hobbes, Thomas. 2010. *Behemoth.* Oxford: Clarendon Press.

Hobbes, Thomas. 2012. *Leviathan.* Noel Malcolm, ed. Oxford: Clarendon Press. First published 1651 and translated into Latin in 1668; cited with 2012 and 1651/1668 pagination.

Kepler, Johannes. 1606. *Ad Vitellionem Paralipomena Quibus Astronomiae Pars Optica Traditur.* Frankfurt: Claudium Marnium and Haeredes Joannis Aubrii.

Leibniz, G W. 1976. *Philosophical Papers and Letters.* Leroy E. Loemaker, ed. Dordrecht: Springer.

Locke, John. 1975. *An Essay Concerning Human Understanding.* Peter H. Nidditch, ed. Oxford: Clarendon.

Mill, John Stuart. 1974. *A System of Logic Ratiocinative and Inductive.* Books 1–III. J. M. Robson, ed. Vol. VII in *Collected Works of John Stuart Mill.* Toronto: University of Toronto Press.

Savile, Henry. 1621. *Praelectiones Tresdecim in Principium Elementorum Euclidis.* Oxford: Excudebant Iohannes Lichfield, & Iacobvs Short.

Wallis, John. 1657. *Mathesis Universalis,* in *Operum Mathematicorum pars prima.* Oxford: Typis *Leonardi Lichfield* Academiæ Typographi. Impensis *Tho. Robinson.*

Wallis, John. 1662. *Hobbius Heauton-timorumenos, or, A Consideration of Mr Hobbes his Dialogues.* Oxford: Printed by A. & L. Litchfield.

Secondary Works

Abizadeh, Arash. 2015. "The Absence of Reference in Hobbes' Philosophy of Language." *Philosophers' Imprint* 15.22: 1–17.

Abizadeh, Arash. 2017. "Hobbes on Mind: Practical Deliberation, Reasoning, and Language." *Journal of the History of Philosophy* 55.1: 1–34.

Abizadeh, Arash. 2018. *Hobbes and the Two Faces of Ethics.* Cambridge: Cambridge University Press.

Adams, Marcus P. 2009. "Empirical Evidence and the Knowledge-that/Knowledge-how Distinction." *Synthese* 170: 97–114.

Adams, Marcus P. 2014a. "The Wax and the Mechanical Mind: Reexamining Hobbes's Objections to Descartes's *Meditations.*" *British Journal for the History of Philosophy* 22: 403–424.

Adams, Marcus P. 2014b. "Hobbes, Definitions, and Simplest Conceptions." *Hobbes Studies* 27: 35–60.

Adams, Marcus P. 2016a. "Hobbes on Natural Philosophy as 'True Physics' and Mixed Mathematics." *Studies in History and Philosophy of Science* 56: 43–51.

Adams, Marcus P. 2016b. "Visual Perception as Patterning: Cavendish against Hobbes on Sensation." *History of Philosophy Quarterly* 33: 193–214.

Adams, Marcus P. 2017. "Natural Philosophy, Deduction, and Geometry in the Hobbes-Boyle Debate." *Hobbes Studies* 30: 83–107.

Adams, Marcus P. 2019. "Hobbes's Laws of Nature in *Leviathan* as a Synthetic Demonstration: Thought Experiments and Knowing the Causes." *Philosophers' Imprint* 19.5: 1–23.

Adams, Marcus P. 2021. "The Presentation and Structure of Thomas Hobbes's Philosophy." In Marcus P. Adams, ed. *A Companion to Hobbes.* Hoboken: Wiley-Blackwell, 1–19.

WORKS CITED 203

Adams, Marcus P. 2022. "Natural Philosophy, Abstraction, and Mathematics among Materialists: Thomas Hobbes and Margaret Cavendish on Light." *Philosophies* 7.2 (special issue on Hobbes's Philosophy of Science edited by Douglas Jesseph): 1–15.

Ashworth, E. J. 1974. *Language and Logic in the Post-Medieval Period.* Dordrecht: Springer.

Baldin, Gregorio. 2020. *Hobbes and Galileo: Method, Matter and the Science of Motion.* Dordrecht: Springer.

Barnouw, Jeffrey. 1980a. "Hobbes's Causal Account of Sensation." *Journal of the History of Philosophy* 18.2: 115–130.

Barnouw, Jeffrey. 1980b. "Vico and the Continuity of Science: The Relation of His Epistemology to Bacon and Hobbes." *Isis* 71.4: 609–620.

Biener, Zvi. 2016. "Hobbes on the Order of Sciences: A Partial Defense of the Mathematization Thesis." *The Southern Journal of Philosophy* 54.3: 312–332.

Boonin-Vail, David. 1994. *Thomas Hobbes and the Science of Moral Virtue.* Cambridge: Cambridge University Press.

Brandt, Frithiof. 1928. *Thomas Hobbes' Mechanical Conception of Nature.* Copenhagen: Levin & Munksgaard.

Brito Vieira, Mónica. 2009. *The Elements of Representation in Hobbes: Aesthetics, Theatre, Law, and Theology in the Construction of Hobbes's Theory of the State.* Leiden: Brill.

Brito Vieira, Mónica. 2021. "Hobbesian Persons and Representation." In Marcus P. Adams, ed., *A Companion to Hobbes.* Hoboken: Wiley-Blackwell, 187–202.

Brown, Gary I. 1991. "The Evolution of the Term 'Mixed Mathematics.'" *Journal of the History of Ideas* 52.1: 81–102.

Chadwick, Alexandra. 2020. "From Soul to Mind in Hobbes's *The Elements of Law.*" *History of European Ideas* 46.3: 257–275.

Dawson, Hannah. 2007. *Locke, Language, and Early Modern Philosophy.* Cambridge: Cambridge University Press.

Dear, Peter. 1988. *Mersenne and the Learning of the Schools.* Ithaca, NY: Cornell University Press.

de Jong, W.R. 1986. "Hobbes's Logic: Language and Scientific Method." *History and Philosophy of Logic* 7.2: 124–142.

de Jong, W.R. 1990. "Did Hobbes have a Semantic Theory of Truth?" *Journal of the History of Philosophy* 28.1: 63–88.

Deigh, John. 1996. "Reason and Ethics in Hobbes's *Leviathan.*" *Journal of the History of Philosophy* 34.1: 33–60.

Deigh, John. 2003. "Reply to Mark Murphy." *Journal of the History of Philosophy* 41.1: 97–109.

Duncan, Stewart. 2011. "Hobbes, Signification, and Insignificant Names." Hobbes Studies 24.2: 158–178.

Duncan, Stewart. 2016. "Hobbes on Language: Propositions, Truth, and Absurdity." In A.P. Martinich and Kinch Hoekstra, eds., *The Oxford Handbook of Hobbes.* Oxford: Oxford University Press, 57–72.

Duncan, Stewart. 2017. "Hobbes, Universal Names, Nominalism." In Stefano Di Bella and Tad M. Schmaltz, eds., *The Problem of Universals in Early Modern Philosophy.* Oxford: Oxford University Press, 41–61.

Evrigenis, Ioannis. 2006. "Hobbes's Thucydides." *Journal of Military Ethics* 5.4: 303–316.

Evrigenis, Ioannis. 2016. "The State of Nature." In A.P. Martinich and Kinch Hoekstra, eds., *The Oxford Handbook of Hobbes.* Oxford: Oxford University Press, 221–241.

Foisneau, Luc. 2021. "Against Philosophical Darkness: A Political Conception of Enlightenment." In Marcus P. Adams, ed., *A Companion to Hobbes.* Hoboken: Wiley-Blackwell, 271–286.

Gauthier, David. 1997. "Hobbes on Demonstration and Construction." *Journal of the History of Philosophy* 35: 509–521.

Gauthier, David. 2001. "Hobbes: The Laws of Nature." *Pacific Philosophical Quarterly* 82: 258–284.

Gaukroger, Stephen. 1986. "Vico and the Maker's Knowledge Principle." *History of Philosophy Quarterly* 3.1: 29–44.

204 WORKS CITED

Giudice, Franco. 2016. "The Most Curious of Sciences: Hobbes's Optics." In A.P. Martinich and Kinch Hoekstra, eds., *The Oxford Handbook of Hobbes*. Oxford: Oxford University Press, 149–168.

Goldenbaum, Ursula. 2008. "Indivisibilia Vera—How Leibniz Came to Love Mathematics." In Ursula Goldenbaum and Douglas Jesseph, eds., *Infinitesimal Differences: Controversies between Leibniz and His Contemporaries*. New York: De Gruyter, 53–94.

Gorham, Geoffrey. 2014. "Hobbes on the Reality of Time." *Hobbes Studies* 27: 80–103.

Grene, Marjorie. 1969. "Hobbes and the Modern Mind: An Introduction." In Marjorie Grene, ed. *The Anatomy of Knowledge*. London: Routledge, 1–30.

Guyer, Paul and Rolf-Peter Horstmann. 2015. "Idealism." In Edward N. Zalta, ed., *The Stanford Encyclopedia of Philosophy*. Available at: https://plato.stanford.edu/archives/fall2020/entries/idealism/

Hankinson, R.J. 2005. "Aristotle on Kind-Crossing." In R.W. Sharples, ed., *Philosophy and the Sciences in Antiquity*. Aldershot: Ashgate, 23–54.

Hampton, Jean. 1986. *Hobbes and the Social Contract Tradition*. Cambridge: Cambridge University Press.

Hattab, Helen. 2014. Hobbes's and Zabarella's Methods: A Missing Link." *Journal of the History of Philosophy* 52.3: 461–486.

Hattab, Helen. 2021. "Hobbes's Unified Method for *Scientia*." In Marcus P. Adams, ed., *A Companion to Hobbes*. Hoboken: Wiley-Blackwell, 25–44.

Hintikka, Jaako and Unto Remes. 1974. *The Method of Analysis: Its Geometrical Origin and Its General Significance*. Dordrecht: D. Reidel Publishing Co.

Heinrichs, T.A. 1973. "Language and Mind in Hobbes." *Yale French Studies* 49: 56–70.

Henry, John. 2016. "Hobbes, Galileo, and the Physics of Simple Circular Motions." *Hobbes Studies* 29: 9–38.

Herbert, Gary B. 1987. "Hobbes's Phenomenology of Space." *Journal of the History of Ideas* 48.4: 709–717.

Hobbes, Thomas. 1660. *Examinatio et Emendatio*. London: Andrew Crooke. Reprinted in OL IV.

Hoekstra, Kinch. 2003. "Hobbes on Law, Nature, and Reason." *Journal of the History of Philosophy* 41.1: 111–120.

Hoekstra, Kinch. 2004. "Disarming the Prophets: Thomas Hobbes and Predictive Power." *Rivista di Storia Della Filosofia* 1: 97–153.

Horstmann, F. 2001. "Hobbes on Hypotheses in Natural Philosophy." *The Monist* 84.4: 487–501.

Hungerland, I.C. and G.R. Vick. 1981. "Hobbes's Theory of Language, Speech, and Reasoning." In I.C. Hungerland and G.R. Vick, eds., *Computatio Sive Logica*. Norwalk, CT: Abaris Books, 7–169.

Jesseph, Douglas. 1993. "Of Analytics and Indivisibles: Hobbes on the Methods of Modern Mathematics." *Revue d'historie des sciences* 46: 153–193.

Jesseph, Douglas. 1996. "Hobbes and the Method of Natural Science." In Tom Sorell, ed., *The Cambridge Companion to Hobbes*. Cambridge: Cambridge University Press, 86–107.

Jesseph, Douglas. 1999. *Squaring the Circle: The War between Hobbes and Wallis*. Chicago: Chicago University Press.

Jesseph, Douglas. 2004. "Galileo, Hobbes, and the Book of Nature." *Perspectives on Science* 12.2: 191–211.

Jesseph, Douglas. 2006. "Hobbesian Mechanics." In Daniel Garber and Steven Nadler, eds., *Oxford Studies in Early Modern Philosophy* III. Oxford: Clarendon Press, 119–152.

Jesseph, Douglas. 2008. "Truth in Fiction: Origins and Consequences of Leibniz's Doctrine of Infinitesimal Magnitudes." In Ursula Goldenbaum and Douglas Jesseph, eds., *Infinitesimal Differences: Controversies between Leibniz and his Contemporaries*. New York: De Gruyter, 215–234.

Jesseph, Douglas. 2009. "*Scientia* in Hobbes." In T. Sorell, G.A. Rogers, and J. Kraye, eds., Scientia in Early Modern Philosophy: Seventeenth-Century Thinkers on Demonstrative Knowledge from First Principles. Dordrecht: Springer, 117–128.

WORKS CITED   205

Jesseph, Douglas. 2015. "Hobbes's Theory of Space." In Vincenzo De Risi, ed., *Mathematizing Space: The Objects of Geometry from Antiquity to the Early Modern Age*. Dordrecht: Springer, 193–208.

Jesseph, Douglas. 2016. "Hobbes on the Foundations of Natural Philosophy." In A.P. Martinich and Kinch Hoekstra, eds., *The Oxford Handbook of Hobbes*. Oxford: Oxford University Press, 134–148.

Jesseph, Douglas. 2017. "Hobbes on the Ratios of Motions and Magnitudes: The Central Task of *De Corpore*, Part III." *Hobbes Studies* 30.1: 58–82.

Jesseph, Douglas. 2021. "Hobbesian Mathematics and the Dispute with Wallis." In Marcus P. Adams, ed., *A Companion to Hobbes*. Hoboken, NJ: John Wiley and Sons, 57–74.

Koethe, J. 2002. "Comments and Criticism: Stanley and Williamson on Knowing How." *Journal of Philosophy* 99.6: 325–328.

Kraynak, Robert. 1982. "Hobbes's Behemoth and the Argument for Absolutism." *The American Political Science Review* 76.4: 837–847.

Kraynak, Robert. 1990. *History and Modernity in the Thought of Thomas Hobbes*. Ithaca: Cornell University Press.

Laird, John. 1968. *Hobbes*. New York: Russell & Russell. Originally published 1934.

Lear, Jonathan. 1982. "Aristotle's Philosophy of Mathematics." *Philosophical Review* 91.2: 161–192.

Leijenhorst, Cees. 1996. "Hobbes's Theory of Causality and its Aristotelian Background." *The Monist* 79.3: 426–447.

Leijenhorst, Cees. 2002. *The Mechanisation of Aristotelianism*. Leiden: Brill.

Lennox, James G. 1986. "Aristotle, Galileo, and the 'Mixed Sciences'." In William A. Wallace, ed., *Reinterpreting Galileo*. Washington, DC: Catholic University Press, 29–51.

Lindberg, David C. 1978. "The Science of Optics." In David C. Lindberg, ed. *Science in the Middle Ages*. Chicago: University of Chicago Press, 338–368.

Lloyd, S.A. 2009. *Morality in the Philosophy of Thomas Hobbes: Cases in the Law of Nature*. Cambridge: Cambridge University Press.

Lund, William. 2003. "Neither *Behemoth* nor *Leviathan*: Explaining Hobbes's Illiberal Politics." *Filozofski vestnik* XXIV.2: 59–83.

Machamer, Peter. 1978. "Galileo and the Causes." In Robert E. Butts and Joseph C. Pitt, eds., *New Perspectives on Galileo*. Dordrecht: D. Reidel Publishing, 161–180.

MacDonald Ross, George. 1987. "Hobbes's Two Theories of Meaning." In A.E. Benjamin, G.N. Cantor and J.R.R. Christie, eds., *The Figural and the Literal: Problems of Language in the History of Science and Philosophy, 1630–1800*. Manchester: Manchester University Press.

MacGillivray, Royce. 1970. "Thomas Hobbes's History of the English Civil War: A Study of *Behemoth*." *Journal of the History of Ideas* 31.2: 179–198

Mahoney, Michael S. 1990. "Barrow's Mathematics: Between Ancients and Moderns. In Mordechai Feingold, ed., *Before Newton: The Life and Times of Issac Barrow*. Cambridge: Cambridge University Press, 179–249.

Malcolm, Noel. 1996. "A Summary Biography of Hobbes." In T. Sorell, ed., *The Cambridge Companion to Hobbes*. Cambridge: Cambridge University Press, 13–44.

Malcolm, Noel. 2002. *Aspects of Hobbes*. Oxford: Oxford University Press.

Malet, Antoni. 1997. "Isaac Barrow on the Mathematization of Nature: Theological Voluntarism and the Rise of Geometrical Optics." *Journal of the History of Ideas* 58.2: 265–287.

Malet, Antoni. 2001. "The Power of Images: Mathematics and Metaphysics in Hobbes's Optics. *Studies in History and Philosophy of Science* 32.2: 303–333.

Mancuso, Paolo. 1996. *Philosophy of Mathematics & Mathematical Practice in the Seventeenth Century*. Oxford: Oxford University Press.

Martin, R.M. 1953. "On the Semantics of Hobbes." *Philosophy and Phenomenological Research* 14: 205–211.

Martinich, A.P. 1999. *Hobbes: A Biography*. Cambridge: Cambridge University Press.

Martinich, A.P. 2005. *Hobbes*. New York: Routledge.

Martinich, A.P. 2010. "Reason and Reciprocity in Hobbes's Political Philosophy: On Sharon Lloyd's *Morality in the Philosophy of Thomas Hobbes*." *Hobbes Studies* 33: 158–169.

206 WORKS CITED

Martinich, A.P. 2019. "Hobbes's Political-Philosophical Project: Science and Subversion." In S. A. Lloyd, ed., *Interpreting Hobbes's Political Philosophy*. Cambridge: Cambridge University Press, 29–49.

Martinich, A.P. 2002. *The Two Gods of the* Leviathan: *Thomas Hobbes on Religion and Politics*. Cambridge: Cambridge University Press.

Martinich, A. P. and Kinch Hoekstra, eds. 2016. *The Oxford Handbook of Hobbes*. Oxford: Oxford University Press.

McKirahan, Richard D. 1978. "Aristotle's Subordinate Sciences." *British Journal for the History of Science* 11: 197–220.

McIntyre, R.W. 2021. "'A Most Useful Economy': Hobbes on Linguistic Meaning and Understanding." In Marcus P. Adams, ed., *A Companion to Hobbes*. Hoboken: Wiley-Blackwell, 93–108.

McNeilly, F.S. 1968. *The Anatomy of* Leviathan. New York: St. Martin's Press.

Médina, José. 2016. "Hobbes's Geometrical Optics." *Hobbes Studies* 27.1: 39–65.

Miller, Ted H. 2011. *Mortal Gods: Science, Politics, and the Humanist Ambitions of Thomas Hobbes*. State College: Penn State University Press.

Mintz, Samuel. 1952. "Galileo, Hobbes, and the Circle of Perfection." *Isis* 43: 98–100.

Moffett, M. and J. Bengson. 2007. "Know-How and Concept Possession." *Philosophical Studies* 136: 31–57.

Murphy, Mark. 2000. "Desire and Ethics in Hobbes's *Leviathan*." *Journal of the History of Philosophy* 38.2: 259–268.

Nauta, Lodi. 2012. "Anti-Essentialism and the Rhetoricization of Knowledge: Mario Nizolio's Humanist Attack on Universals." *Renaissance Quarterly* 65.1: 31–66.

Normore, Calvin. 2017. "Nominalism." In Henrik Lagerland and Benjamin Hill, eds., *Routledge Companion to Sixteenth Century Philosophy*. New York: Routledge, 121–136.

Nuchelmans, Gabriel. 1980. *Late-Scholastic and Humanist Theories of the Proposition*. Amsterdam: North Holland Publishing Co.

Nuchelmans, Gabriel. 1983. *Judgement and Proposition: From Descartes to Kant*. Amsterdam: North Holland Publishing Co.

Noë, A. 2005. "Against Intellectualism." *Analysis* 65.4: 278–290.

Otte, Michael and Marco Panza. 1997. *Analysis and Synthesis in Mathematics: History and Philosophy*. Dordrecht: Kluwer.

Pacchi, A. 1965. *Convenzione e ipotesi nella formazione della filosophia natural di Thomas Hobbes*. Firenze: La Nuova Italia (Pubblicazioni della Facoltà di Lettere e Filosofia dell'Università degli Studi di Milano, 38).

Paganini, Gianni. 2019. "Hobbes's Philosophical Method and the Passion of Curiosity." In S.A. Lloyd, ed., *Interpreting Hobbes's Political Philosophy*. Cambridge: Cambridge University Press, 50–69.

Pavese, Carlotta. 2021. "Knowledge How." In Edward N. Zalta, ed., *The Stanford Encyclopedia of Philosophy* (Summer 2021 Edition). Available at: https://plato.stanford.edu/entries/knowle dge-how/

Pécharman, Martine. 2016. "Hobbes on Logic, or How to Deal with Aristotle's Legacy." In A.P. Martinich and Kinch Hoekstra, eds., *The Oxford Handbook of Hobbes*. Oxford: Oxford University Press, 21–59.

Pérez-Ramos, A. 1989. *Francis Bacon's Idea of Science and the Maker's Knowledge Tradition*. Oxford: Oxford University Press.

Peters, Richard. 1956. *Hobbes*. Baltimore: Penguin Books.

Pettit, Philip. 2008. *Made with Words: Hobbes on Language, Mind, and Politics*. Princeton: Princeton University Press.

Prins, Jan. 1987. "Kepler, Hobbes, and Medieval Optics." *Philosophia Naturalis* 24: 287–310.

Prins, Jan. 1993. "Ward's Polemic with Hobbes on the Sources of His Optical Theories." *Revue d'histoire des sciences* 46: 195–224.

Prins, Jan. 1996. "Hobbes on Light and Vision." In Tom Sorell, ed., *The Cambridge Companion to Hobbes*. Cambridge: Cambridge University Press, 129–159.

WORKS CITED 207

Rampelt, Jason. 2019. *Distinctions of Reason and Reasonable Distinctions: The Academic Life of John Wallis (1616--1703)*. Leiden: Brill.

Raylor, Timothy. 2018. *Philosophy, Rhetoric, and Thomas Hobbes*. Oxford: Oxford University Press.

Reif, Sister Mary Richard. 1962. "Natural Philosophy in Some Early Seventeenth Century Scholastic Textbooks." PhD diss., St. Louis University.

Reif, Sister Patricia. 1969. "The Textbook Tradition in Natural Philosophy." *Journal of the History of Ideas* 30.1: 17–32.

Robertson, George C. 1886. *Hobbes*. London: William Blackwood.

Rossi, M.M. 1942. *Alle fonti del deismo e del materialismo moderno*. Firenze: La Nuova Italia.

Ryan, Alan. *The Philosophy of the Social Sciences*. New York: Pantheon Books, 1970.

Ryle, Gilbert. 1949. *The Concept of Mind*. London: Hutchinson.

Schuhmann, Karl. 1998. "Skinner's Hobbes." *British Journal of the History of Philosophy* 6.1: 115–125.

Schuhmann, Karl. 2000. "Hobbes's Concept of History." In G.A. J. Rogers and Tom Sorell, eds. *Hobbes and History*. London: Routledge, 1–24.

Shapin, Steven and Simon Schaffer. 1985. *Leviathan and the Air-Pump: Hobbes, Boyle, and the Experimental Life*. Princeton: Princeton University Press.

Shapiro, Alan. 1973. "Kinematic Optics: A Study of the Wave Theory of Light in the Seventeenth Century." *Archive for the History of the Exact Sciences* 11: 134–266.

Schwartz, Avshalom. 2020. "The Sleeping Subject: On the Use and Abuse of Imagination in Hobbes's Leviathan." *Hobbes Studies* 33: 153–175.

Skinner, Quentin. 1996. *Reason and Rhetoric in the Philosophy of Hobbes*. Cambridge: Cambridge University Press.

Skinner, Quentin. 2002. *Visions of Politics. Volume III: Hobbes and Civil Science*. Cambridge: Cambridge University Press.

Slomp, Gabriella. 2000. *Thomas Hobbes and the Political Philosophy of Glory*. London: Palgrave Macmillan.

Slowik, Edward. 2014. "Hobbes and the Phantasm of Space." *Hobbes Studies* 27.1: 61–79.

Slowik, Edward. 2021. "Body and Space in Hobbes and Descartes." In Marcus P. Adams, ed., *A Companion to Hobbes*. Hoboken: Wiley-Blackwell, 367–380.

Snowdon, P. 2003. "Knowing How and Knowing That: A Distinction Reconsidered." *Proceedings of the Aristotelian Society* 104.1: 1–29.

Soles, Deborah Hansen. 1996. *Strong Wits and Spider Webs: A Study of Hobbes's Philosophy of Language*. Aldershot: Avebury.

Sorell, Tom. 1986. *Hobbes*. New York: Routledge & Kegan Paul.

Sorell, Tom. 1995. "Hobbes's Objections and Hobbes's System." In Roger Ariew and Marjorie Grene, eds., *Descartes and His Contemporaries*. Chicago: University of Chicago Press, 83–96.

Sorell, Tom. 2019. "Appeals to Experience in Hobbes's Science of Politics." In Alberto Vanzo and Peter R. Anstey, eds., *Experiment, Speculation, and Religion in Early Modern Philosophy*. New York: Routledge, 81–100.

Sowerby, Robin. 1998. "Thomas Hobbes's Translation of Thucydides." *Translation and Literature* 7.2: 147–169.

Springborg, Patricia. 1994. "Hobbes, Heresy, and the *Historia Ecciesiastica*." *Journal of the History of Ideas* 55.4: 553–571.

Springborg, Patricia. 2003. "*Behemoth* and Hobbes's 'Science of Just and Unjust.'" *Filozofski vestnik* XXIV.2: 267–289.

Stanley, Jason and Timothy Williamson. 2001. "Knowing How." *Journal of Philosophy* 98.8: 411–444.

Stauffer, Devin. 2018. *Hobbes's Kingdom of Light*. Chicago: University of Chicago Press.

Steinmetz, Alicia. 2021. "Hobbes and the Politics of Translation." *Political Theory* 49.1: 83–108.

Tabb, Kathryn. 2014. "The Fate of Nebuchadnezzar: Curiosity and Human Nature in Hobbes." *Hobbes Studies* 27: 13–34.

208 WORKS CITED

Talaska, Richard. 1988. "Analytic and Synthetic Method According to Hobbes." *Journal of the History of Philosophy* 26.2: 207–237.

Taylor, A.E. 1938. "The Ethical Doctrine of Hobbes." *Philosophy* 13: 406–424.

Tuck, Richard. 1988. "Hobbes and Descartes." In G.A.J. Rogers and Alan Ryan, eds., *Perspectives on Thomas Hobbes*. Oxford: Clarendon Press, 11–41.

van Apeldoorn, Laurens. 2012. "Reconsidering Hobbes's Account of Practical Deliberation." *Hobbes Studies* 25: 143–165.

Vaughn, Geoffrey. 2002. Behemoth *Teaches* Leviathan. Lanham: Lexington Books.

Vaughn, Geoffrey. 2009. "The Audiences of '*Behemoth*' and the Politics of Conversation." In Tomaz Mastnak, ed., *Hobbes's Behemoth: Religion and Democracy*. Exeter: Imprint Academic, 170–185.

Wallace, William. 1991. *Galileo, the Jesuits and the Medieval Aristotle*. Aldershot, Hampshire, UK: Variorum.

Warrender, Howard. 1957. *The Political Philosophy of Hobbes*. Oxford: Oxford University Press.

Watkins, J.W.N. 1973. *Hobbes's System of Ideas*. London: Hutchison.

Wilson, Fred. 1996. "Hobbes' Inductive Methodology." *History of Philosophy Quarterly* 13.2: 167–186.

Index

For the benefit of digital users, indexed terms that span two pages (e.g., 52–53) may, on occasion, appear on only one of those pages.

Abizadeh, Arash 25–26, 27n.26, 34, 35n.36,
  58–59n.17, 70, 71n.37, 86–87
Adam, as a lone philosopher 34–35, 36–44,
  40–41n.43, 47, 79, 86, 100, 113
air, nature of 153–59
analysis and synthesis 2–3n.3, 41–42, 65–77,
  96–100, 113, 116, 178n.11
Aristotle 2, 31n.32, 50, 69n.35, 79–80n.45,
  81–83n.49, 94, 102, 104, 130n.3, 131–32,
  135–36, 138, 140, 188, 189
  see also mixed mathematics: Aristotle on
    (subalternate sciences)
Aubrey, John 1, 5–6, 13, 122

baptism, sacrament of 195–96
Barrow, Isaac 130–32, 134, 135, 138, 145n.19,
  147–48, 149–50, 160
Berkeley, George 62–63, 62–63n.23, 77n.42
body 2–4, 60–62, 63–65, 68–70, 72–73, 88–93,
  95, 100–2, 104–5, 107–11, 116, 132n.6,
  134–38, 141–44, 147–51, 154–56, 161–64,
  167, 169
  accidents of 21n.14, 44
  definition of 108–9
  fluid, nature of 148, 149–51, 153, 163–64
  individuation of 37–38
  macroscopic versus microscopic 5
  see also conceptions; simplest
Boyle, Robert 8, 12–13, 130–31, 136–37,
  138n.14, 151–67, 170n.1, 179–80, 199
Brandt, Frithiof 62–63n.23, 77n.42, 145n.19

cause
  as entire cause 73–76, 117, 180–81, 186, 198
  as possible cause 129–30, 134, 137–38, 139,
    140–41, 150–52, 157, 159–60
  see also Knowledge, scientific (*scientia*);
    Maker's knowledge; Motion: in
    definitions
Charles I, King 29n.28, 180–81, 186
Church, Roman Catholic 181–82, 188, 190,
  194–96, 197–98 *see also* Pope
civil philosophy 66n.26, 112, 122–25, 178–79
  beginning of 89–92, 110–14
  as beginning with Hobbes 57n.15, 94

and eloquence 13–15
and history 169–71, 175–77
objects of 75
as scientific knowledge 3, 5–7, 9–10, 12, 50,
  51–52, 53–54, 57, 75–76, 88, 96–97
use in historiography 181, 183–98, 200
  see also Maker's knowledge
commonwealth 7, 10, 176–77, 183–85, 187–88
  construction of 3, 53–54, 57, 81–83n.49, 96,
    97, 121n.31, 126–28
conceptions
  decay of 20, 24–33, 40, 56, 64–65, 110–11
  and error 27–28, 32n.33, 32–33, 70
  as images 16, 19–24, 27–28, 35–36, 39, 41–42,
    60–61, 62–63n.23, 63, 68–69, 70, 92–93,
    107–8, 144n.18, 196
  and motions 23
  reception of 22–23, 38–39, 66n.26
  as representing all 62–63n.23, 86
  simplest/simple 2–3n.3, 3–4, 20, 69–70,
    72–73, 88–89, 95–98, 100–1, 101n.9, 102,
    104–5, 109–11, 113–14, 115n.27, 120,
    121n.31, 121–22, 124, 125, 129n.1, 135–36,
    167, 177n.10, 177
  in trains of imaginations 28–31, 29–30n.30,
    36–37, 48–49, 53, 59–60
considering as 2, 9, 41–42, 56–71, 58n.16, 77–78n.43,
    85, 86n.54, 92–93, 95–96, 105–6, 115n.27,
    123, 135–36, 161, 199–200
  versus abstraction 62–64
  and Aristotle on mathematical objects 131
  and curiosity 64–66
  versus dividing 60–62, 67–68
contemplation *see* Hobbes: on meditating/
    meditation
contract (covenant) 34, 119, 127
convention 34–35, 102, 172n.3
curiosity 14, 29n.29, 29–30n.30, 33–34, 93–94,
    191–92
  and language, 34, 40n.42, 59–60n.19
  and non-human animals 59n.18
  and scientific knowledge 50, 51, 57, 57n.12, 58,
    58n.16, 59–60, 59n.18, 59–60n.19, 64–66,
    81–83n.49, 83–84, 93–94, 135–36, 187–88
  see also considering as: and curiosity

210   INDEX

definitions *see* language: definitions; motion: in
    definitions
demons, belief in 196–97
demonstration 2–7, 9–12, 14–15, 50, 55, 66–67,
    67n.28, 73, 95–106, 111–12, 115–28, 129,
    178–79
    that versus why 11–12, 31, 129, 131, 133, 139–
        40, 141–44, 147–52, 159, 166–68, 170–71,
        175–77, 179–81, 190–98
    see also language: and demonstration;
        language: and syllogisms; Laws of
        Nature: demonstration (as cause of
        peace)
Descartes, René 16n.1, 18–19, 39–40, 62–63n.23,
    65, 144n.18, 177n.10
    versus Hobbes on dreams 17–18
    versus Hobbes on skeptical worries 16–17n.3,
        16–18
    on the idea of the sun 92–93

endeavor 6–7, 10–11, 110–11, 117–18, 133, 141,
    142–44, 144n.18, 147, 150, 154–55, 157,
    160–61
equality, in state of nature 110–12, 118, 119–20,
    190–91
eucharist 195–96
Euclid 95, 103–4, 122
    influence on Hobbes 1, 4
    see also Hobbes, Thomas: criticisms of Euclid
experience/experiments 129, 133, 142, 144, 148,
    151–52, 153–59, 166–68, 169, 172, 189
    confirmation by 112–13, 171, 177–79
    as unsuitable for scientific knowledge 128,
        165, 179

geometry
    as abstract 64
    versus arithmetic 131–32
    objects of 2–3, 53n.7, 65, 66–67n.27, 67n.30,
        75, 92, 115n.27
        triangle 77–87
    as scientific knowledge 50, 51–52, 53–54
    see also Considering as; Knowledge, scientific
        (*scientia*); Maker's knowledge
God 16–17n.3, 28n.27, 34, 107n.16, 153, 182–83,
    192–93
    as imitated in philosophizing 3, 9, 56, 56n.11,
        81–83n.49
    as judge 185
    Kingdom of 194–95
    as knowing causes of natural phenomena 34–35,
        55–56, 134
    no conception of 104, 104n.15
    versus polytheistic gods 192–93

Hampton, Jean 5, 6–7, 89, 129n.2
Hattab, Helen, 3–4n.4, 79–80n.45, 95n.3,
    101n.9, 133n.8
heat, explanation of 147–51
history 169–70, 172–77
    civil 8, 12, 133, 169, 175
    Hobbesian historiography 12, 170–71,
        198–200
        as mixed 194
        and suppositional certainty 12, 194
    judgment in 173–74, 173n.5
    limits of 173, 174
    natural 8, 133, 138–39, 151, 165, 169–70, 175
    praise of 173–74
    see also philosophy: subject of
Hobbes, Thomas
    on dreams 17–18, 24n.18, 24–25
    on English Civil Wars 179–90
    Euclid, criticisms of 2–4, 55n.8, 67n.28, 73,
        73n.40, 95, 115, 122, 133, 135–36
    as idealist 16
    on light 13
    on Meditating/meditation 71n.38, 72–73,
        78–79, 80, 80n.46, 81–83, 81–83n.49, 104,
        186–87
    on skepticism, 16–18
    Table of the Several Subjects of Science 6n.6, 119
    see also Maker's knowledge; memory: motion;
        philosophy: deductivist view of Hobbes's
        Philosophy; language: definitions
Hume, David, 53, 176–77

ideas *see* conceptions
imagination 24, 24n.18
    and trains of imaginations, 28–30
    see also conceptions

Jesseph, Douglas 11n.7, 16n.2, 58n.16, 73n.40,
    77n.42, 83n.50, 94, 103n.13, 103–4, 131–
    32, 132n.6, 140, 143, 152n.27
judgment,
    faculty of 21–22n.15
    private 182–86
        and Original Sin 185
    of similitudes in prudence 18, 25–26, 28–29,
        33, 36, 37–44, 43n.45, 56–57, 58–59, 64–
        65, 70–71, 73–74, 81, 84–85
    see also history: judgment in
justice, 13, 53–54, 81–83n.49, 114, 117–21, 123,
    170n.1, 188–90

Kant, Immanuel 16
knowledge, scientific (*scientia*), 2–3, 9, 15, 18,
    31, 50–51, 133

INDEX 211

versus apprehending 2–3n.3, 69–70, 115, 115n.26
as causal 50, 52–56, 57n.12
criteria for 50, 87–88
discovery of (method of discovery) 77–87, 79–80n.45, 113n.24, 124
early account in *Elements of Law* 172
as founded in a particular conception 58–59n.17, 62–63, 66–67, 70–71, 73–74, 76–77, 77n.42, 81–83n.49, 93–94
and invention 20n.12, 78–81, 79–80n.45, 113n.24
as power 37, 51–52, 87–88
and power versus eloquence 14, 52n.4
versus prudence 57, 59–60, 73–74
see also language: and scientific knowledge; Maker's knowledge; perspectival principle; prudence

language
definitions 2–4, 54, 74
by explication, 95–96, 97–98, 100–6
see also Motion in definitions (generative definitions)
and demonstration 99
and marks 34–36
names 44–45
abstract 44–45, 59–60, 62–63, 68n.32, 73–74
and explication 100
and scientific knowledge 77, 87
as marks versus signs, 47–48, 95–96
simple versus complex 38–39
nominalism 43n.45, 52–53n.6, 87, 199–200
origin, 34
propositions 40–45
contingent 42–45, 48–49
necessary versus contingent propositions 46–47n.47, 52–53n.6
necessary versus hypothetical 46–47n.47, 52
and prudence 33–48, 56, 73–74
relationship to mind 36n.37, 39–40, 70
and scientific knowledge 51, 73–74, 77–87
special uses of 71–72, 71n.37, 78, 95–96, 116
syllogisms 45–48, 51, 77–78n.43, 98–99, 116–17, 120–21, 124–25, 167
Laws of Nature 5–6, 12, 53–54, 95–96, 117–18, 175–76
and balance 120
demonstration (as cause of peace) 117–21, 186
as an easy sum 15, 120–21, 186–88, 189
as God's commands 123–24
as immutable and eternal 123

Leibniz, G.W., 43n.45, 58n.16
Leijenhorst, Cees 16n.2, 17n.4, 19n.8, 21n.13, 21–22n.15, 61n.20, 106–7, 109n.21, 132n.7
light, 145, 147–48
in *Elements* versus *De corpore* 147
Lloyd, S.A. 95n.2, 96, 124
Locke, John 76

Maker's knowledge 8–13, 14–15
in civil philosophy 117–21, 177
in geometry 115–17, 159–60, 164–65, 168
principle 50, 53–56
types of making 71, 95–96
see also Knowledge, scientific
Malcolm, Noel 6–7, 19n.9, 54–55, 55n.9, 75–76, 85–86n.52, 89, 118, 197n.21
Martinich, A.P. 1n.1, 2–3n.3, 3–4n.4, 4–6, 70n.36, 79n.44, 79–80n.45, 96, 98n.5, 100n.8, 124, 129, 172n.3
memory 18–19, 21–22n.15, 24–26, 27n.26, 30–33, 40, 52, 65, 89–90, 125, 133, 140–41, 172, 177–78
and experience 25–26
see also language: origin
Mill, J.S. 27n.25, 43n.45, 45n.46, 87
mixed mathematics 11–13, 53n.7, 129–31, 133–38, 167–68
Aristotle on (subalternate sciences) 131
Hobbes versus Wallis, 11, 138–40
origin of 130n.3
versus pure 132n.6, 134–36, 155–56
see also Barrow, Isaac; history: Hobbesian historiography; natural philosophy: as mixed mathematics
motion 5, 11–12, 69n.34, 108n.20, 108, 135–36
in definitions (generative definitions) 2–4, 8–9, 54–55, 73, 74, 76–77, 102–3, 115, 119–20, 124, 160
of the mind 6, 16
as mutation 75, 141–43
a priori principles of 24–25, 106–7, 129–31, 134, 136–38, 143, 151, 167–68
simple circular 13, 72n.39, 129–30, 145–51, 148–49n.22, 154–65
as fermentation 132, 147–51, 149n.23, 158, 163–65
as sievelike 161, 162–63
treatment of as mathematical, 132, 164

names *see* language
natural philosophy 3, 52, 55–56, 67n.30, 81–83n.49, 110–11, 129, 159
and historiography 198–200

212 INDEX

natural philosophy (*cont.*)
  Hobbes's as conflicted 54–55
  as mixed mathematics 129, 130–31n.4
  versus natural history 133
  and suppositional certainty 11–12, 76, 129–30,
    140–41, 144, 150–51, 165, 167–68
  suppositions used in, 145
  see also mixed mathematics

optics, 6n.6, 17n.7, 91, 130–31n.4, 131–32, 135,
  144n.18

passions 4–5, 6–7, 10–11, 13–14, 15, 23n.17, 28–
  29, 58n.16, 59–60n.19, 97, 110–13, 114,
  117–20, 125, 177–78
  fear 34, 37, 47–48, 114, 117–18, 190–93
  hope 88, 97–98, 110–14, 117–18, 120, 176,
    190–91, 193
  inclining to peace 88, 111–12, 114, 117–18,
    120, 123
  see also curiosity; endeavor
peace, 9–10, 14–15, 34, 51–52, 53–54, 57n.12,
  64, 88, 89–90, 96–97, 183
  analysis of 121n.31
  construction of 125–28
  definition of 120
  see also Laws of Nature: demonstration (as cause
    of peace); passions: inclining to peace
perspectival principle, 4–5, 37, 51, 75–76,
  88–93, 114
Pettit, Philip 23n.17, 26n.22, 32, 33n.34, 39, 40n.42,
  43n.45, 53n.7, 55n.8, 58–59n.17, 77n.42
philosophy,
  deductivist view of Hobbes's Philosophy 4–6,
    125, 129–31, 151–52, 168
    versus disunity view 1
    and reductionist view 5, 6–7, 51, 89–92
  definitions of, 52n.5, 59–60n.19
  disunity view of Hobbes's Philosophy 6–7 *see*
    also natural philosophy
  and Maker's Knowledge view 8–13, 56
  paths of 52, 59–60n.19, 96–97, 140–41
  subject of 169
  unity of philosophy 129, 167
    relationship between *Leviathan* Parts I and
      II 125–28
    requirements for 8
physics *see* natural philosophy
politics *see* civil philosophy
Pope 181–83, 190, 195–96
  bondage by 190–91, 194–96, 197–98, 200
prudence 2–3, 9, 14–15
  and anxiety 191–93, 197–98

as *cognitio* 2–3, 8, 9, 18, 48–49, 50, 81
development of 26–28, 53, 58–59, 73–74
and knowledge of causes 29–30n.30, 52
limitations of, 28, 31, 32–33, 52, 169, 188–89
versus mixed mathematics 130n.3, 168
shared by humans and non-human
  animals 29–30, 33, 191
versus superstition 191, 193
usefulness of, 31
see also language: and prudence

Raylor, Timothy 27, 79–80n.45, 130n.3, 173,
  174, 174n.6, 181n.14
religion, cause of 190–93

Schuhmann, Karl 100n.8, 143n.17, 148–49n.22,
  149n.23, 151n.25, 174–75, 175n.7
science
  definition of 54
  nature of *see* knowledge: scientific; natural
    philosophy
sense 18–24
  explanation of, 140–44
Skinner, Quentin 13–15, 79–80n.45, 174n.6
sovereign 178, 180–81
  as judge 184–85, 185n.17
  power, 10, 57, 182
  rights of 126–28, 179
Springborg, Patricia 169–70, 171n.2, 190

thought experiment 3–4, 71–72, 79–80n.45,
  95n.3, 95–106
  annihilation of the world 17, 20n.11, 20, 52–
    53n.6, 75–76, 106–9, 111
  state of nature 75–76, 90, 91, 97, 110–14
Thucydides 169, 171, 173–74, 174n.6, 180–81
Tuck, Richard, 16, 16–17n.3

universals 86n.53
  apprehension of 69–70, 100–1
  as names 36–39, 70, 77n.42
  see also conceptions: simplest
universities, role in education 187–88, 189–90

Wallis, John 103–4, 130–32, 135n.11, 179
  criticisms by Hobbes on pure/mixed
    mathematics 11, 138–40
  demonstration in pure mathematics 139–40
Ward, Seth 77n.42, 81n.48
White, Thomas 36n.37, 39, 60, 62–63n.23, 63
  Criticisms of White by Hobbes 16, 19, 23, 24–25,
    39, 54, 60, 62–63n.23, 63, 66–67n.27, 69n.35,
    74n.41, 102, 107n.16, 107–8, 135, 164